PRAISE FOR *SPELLBOUND*

"By integrating wisdom from psychology, philosophy, history, and storytelling, *Spellbound* provides a unique perspective on how science informs the psychological reality of magic."

—CAROL BERNSTEIN, FORMER AMERICAN PSYCHIATRIC ASSOCIATION PRESIDENT

"We all love stories. In *Spellbound*, Dr. Daniel Lieberman beautifully weaves together stories from literature, philosophy, spirituality, mythology, and history with neuroscience and Jungian and Freudian psychology. Using clear, jargon-free language, he helps us understand how integrating the magical, mysterious, and metaphorical language of the unconscious, with our rational, orderly, and logical conscious minds, can help us to develop our true selves."

—DEBORAH L. FELDHEIM, M.D., PAST PRESIDENT OF THE WASHINGTON BALTIMORE CENTER FOR PSYCHOANALYSIS

SPELL
BOUND

ALSO BY DANIEL Z. LIEBERMAN, MD

The Molecule of More
(coauthored with Michael E. Long)

SPELL BOUND

Modern Science, Ancient Magic, and the Hidden Potential of the Unconscious Mind

DANIEL Z. LIEBERMAN, MD

BenBella Books, Inc.
Dallas, TX

BENBELLA

BenBella Books, Inc.
10440 N. Central Expressway
Suite 800
Dallas, TX 75231
benbellabooks.com
Send feedback to feedback@benbellabooks.com

Printed in the United States of America
10 9 8 7 6 5 4 3 2 1

Library of Congress Control Number: 2022935025
ISBN 9781637741320 (hardcover)
ISBN 9781637741337 (electronic)

Editing by Leah Wilson
Copyediting by James Fraleigh
Proofreading by Marissa Wold Uhina and Amy Zarkos
Indexing by WordCo
Text design and composition by Aaron Edmiston
Cover design by Sarah Avinger
Cover image © Shutterstock / Kristina Vor
Printed by Lake Book Manufacturing

For St. John's College,
where the books of magic are read

CONTENTS

Part III

TRANSCENDENCE

Know thyself.

—Inscription on the Temple of Apollo at Delphi

Introduction

THE UNCONSCIOUS AND THE SUPERNATURAL

There's someone living in your head besides you.

You think you're calling all the shots—that you're in charge of your thoughts, feelings, and choices. You're not.

When it comes to how you think and feel your way through life, at best you're the copilot. At worst, you're along for the ride, at the mercy of a part of your brain that's overwhelmingly powerful but entirely unseen, influential but utterly secret from you.

It's called *the unconscious*: a vast collection of neural circuits working all the time, just out of sight—just out of mind. It determines whether you will be enthusiastic or bored, full of energy or barely able to keep your eyes open. It can make you compassionate or fill you with hatred. It chooses what you desire and who you fall in love with. It can solve seemingly impossible problems with the gift of inspiration, and sometimes it possesses you completely, plunging you into an alien world of mystical experience.

What if we could learn how to recognize the influences of the unconscious, and make it an ally in helping us become who we were meant to be?

And what if the key to doing so was not modern science . . . but a book of fairy tales and a deck of tarot cards?

SCIENCE AND MAGIC

The unconscious has played a prominent—though hidden—role in human culture throughout history. The ancients sought to understand its influence through myths about supernatural creatures: gods, goddesses, demons, and spirits. This way of thinking has fallen out of favor, but the need to work with our dark twin who accompanies us through life is as important as ever.

Our modern understanding of the mind comes from the disciplines of psychology and neuroscience. Because the brain is so much more complex than other organs, these sciences are still in their infancy, and their explanations work best when it comes to simple behaviors that can be tested in a laboratory. Although it may sound strange, a deeper, more sophisticated understanding of unconscious brain functioning can be obtained by combining these modern scientific discoveries with a psychological understanding gleaned from ancient myths.

Drawing on the work of Swiss psychiatrist Carl Gustav (C. G.) Jung, best known for his theory of the collective unconscious, we can unite science and the supernatural to better understand the hidden part of our minds: to discover how it influences our behavior and how we can join forces with it to enrich our lives and make ourselves whole.

Most people don't think about their unconscious mind very much. In fact, the biggest mistake people make is underestimating its importance. You may have seen pictures that illustrate the mind as an iceberg, with the vast majority (the unconscious) floating beneath the surface. The unconscious uses a much larger part of the brain than the conscious mind, and that's why it has such an oversized influence. It's also more primitive. It has its own way of processing information that's alien to conscious modes of thought.

When the unconscious exerts its influence, it can feel like an external force is acting on us. It can throw us into a passion that takes over our behavior and makes us act in irrational, sometimes self-destructive ways. But it can also awaken latent abilities we never suspected we had. That's why the unconscious has historically been conceived as a supernatural force. It's both overwhelmingly powerful and utterly alien to conscious modes of thought. One of Jung's most important insights was that the best source of knowledge about the unconscious comes from ancient stories about the supernatural—stories about magic.

That's not easy to accept. We live in an age of scientific discovery, in which rational investigation into natural processes has banished the superstitions of previous generations in favor of an understanding that gives us unprecedented control over our environment. We no longer need sun gods, tree spirits, or personal demons to explain natural and psychological phenomena. Today, we confidently state that those explanations are of no use to us, because they're false.

And yet, we remain fascinated by ancient traditions of the supernatural. Some of today's most popular stories are about magic. Enthusiastic audiences pour into theaters to enter the worlds of Harry Potter, Middle-earth, and Marvel's superheroes. At first glance it seems odd that magic and the supernatural are still so compelling, long after their descriptions of how the world works have been proven inaccurate. Disproved scientific theories, by contrast, rarely demonstrate this sort of persistence. Few people care about the Rutherford model of the atom, for example. But magic, discredited centuries ago, continues to hold us in its mysterious grip.

The reason stories about magic fascinate us is that they reveal our inner world. The human brain is more than a rational thinking machine. It has darker corners that give rise to the irrational—things outside conscious control. The rational part of the brain uses language and logic to process experiences. But the irrational parts, which are generally hidden from us, don't think with language; they use symbolism that past centuries of thinkers have dubbed *magic*.

The word *magic* can be used in various ways. The first thing that might come to mind is a stage performer pulling a rabbit out of a hat. But that's not really magic; it's just a trick. Alternatively, we might think of magic spells, flying carpets, or divination. We're getting closer, but to understand the hidden parts of the psyche, we have to go back to a more ancient form of magic, perhaps the most ancient form of all: communicating with otherworldly spirits and, sometimes, being possessed by them.

A magic potion needs more than eye of newt and toe of frog. It must be invested with an invisible power before it can make someone fall in love or give them superhuman abilities. Something beyond the physical ingredients needs to be present, something that enters the one who drinks the potion and transforms him. Even an enchanted scroll needs to be inhabited by an invisible spirit before it can be more than parchment and ink. Magic occurs when the natural world is invaded and occupied by entities from the spirit world.

We see this most clearly in ancient accounts of prophets, oracles, and priestesses, who were thought to be portals through which the gods spoke. Sometimes they were believed to be under the direct control of supernatural entities. Ordinary people also experienced the influence of the gods. Roman soldiers charging into battle, heedless of their own safety, believed they were invested with courage by Mars, the god of war. Lovers experiencing the frenzy of erotic passion believed they were possessed by Venus, the goddess of love. Poets were inspired by the divine Muses, and champions gave thanks to the goddess of victory for endowing them with the ability to win. There was widespread belief that invisible entities could occupy the minds of mortals to impose their will on humankind.

The ancients attributed these experiences to supernatural creatures because it felt as if something outside themselves was responsible for the things they did. Love comes unbidden; it can't be summoned at will. The same is true of poetic inspiration. The ancients reasoned that when they achieved something beyond their usual abilities, taking credit for it themselves made no sense. It came from outside, sometimes as a blessing and sometimes as a curse.

Today, we no longer believe in the pervasive influence of supernatural entities. We credit ourselves with having exclusive control over what goes on inside our heads. We tend to think that if we want to change our behavior, all we have to do is put our mind to it. Self-help books feature ten easy steps to achieve discipline, confidence, and success. Do they work? If they did, we'd all be skinny, rich, happy, and kind. The idea that we can take control over our emotions and summon abilities that lie dormant within us doesn't hold up to the cold light of reality. Despite the progress we've made since science and technology replaced belief in magic, we remain at the mercy of forces beyond our control.

Why do we act in self-destructive ways? Why do we do things knowing full well that in a week, a day, maybe even an hour, we'll regret it? What is the source of our mental energy? Where does inspiration come from? Contemporary research tells us that there are some circuits in the brain we're aware of and others that we're not. Those we're aware of we can control. The ones we're not control us. They are the gods within.

That gives us two ways of understanding the human condition: supernatural and scientific. The supernatural approach attributes irrational behaviors to the influence of magical creatures, the spirits and demons of ancient stories. When we're overcome with rage, we feel as if we're no longer in control of our own behavior. We burn bridges, insult those we love, sometimes even smash prized possessions. Later we might say, "I don't know what possessed me to do that." The ancient belief that humans are liable to be taken over by invisible forces is embedded in our very language.

The scientific approach is different. It explains human behavior through neurochemical activity: the actions of proteins, monoamines, and phospholipids. Although this explanation can be understood only by specialists—and even they understand it poorly—we tend to accept the scientific explanation as the only one that's useful. That's a mistake. To fully understand the human condition, we need both.

THE ANCIENT AND THE MODERN

Many phenomena can be understood from more than one perspective. Matter is one example. In 1927, English astrophysicist Sir Arthur Eddington gave a lecture he called "Two Tables." He pointed out that the table of everyday experience is solid. If you knock your head against it, it will hurt. The table of science, however, isn't solid at all. It's mostly empty space, a ghostly haze filled with insubstantial fields of probability. Which way of understanding is better? It depends what you're doing. If you're a scientist studying quantum physics, it's the incomprehensible table of hazy probabilities. But if you're looking for a place to put down your drink, you want the solid piece of wood.

It's the same with human behavior. Neuroscientists advance our understanding of the brain using the materialist model of interactions among atoms and molecules. But the ancient traditions of magic, traditions that have been developed and refined over the course of millennia, provide us with an understanding of the mind that is frequently more useful. Ancient stories of the supernatural may be factually incorrect, but so is the belief that tables are solid. We choose the model that works best. It would be mentally exhausting to constantly remind ourselves that our instinctual belief in the solidity of matter is an illusion brought about by quantum interactions we don't fully understand. The brain evolved to perceive tables as solid objects, so we only think about the true, underlying reality when we happen to read an article about subatomic phenomena. Similarly, rejecting the traditions of magic as a way of accounting for human behavior is contrary to our instinctual ways of understanding the world and ourselves. Magic has played a central role in human culture since prehistory because our brains evolved to think that way.

The fullest understanding of the human condition comes from combining the two models. As we delve deeper into the human psyche, we'll move back and forth between the two models in order to understand how science sheds light on magic and vice versa. It can be jarring at times to jump between the ancient and the modern, the metaphorical and the literal, but it's the best way to understand the brain, which evolved as a peculiar combination of the rational and the mystical. And to guide us, we'll use the work of C. G. Jung, who believed that myths, folklore, and other stories about the supernatural were manifestations of primitive ways of thinking that, despite being hidden beneath a rational veneer, are still with us today.

PEERING INTO THE DARKNESS

Jung spent much of his career studying the psychological meaning of magical traditions. Drawing on analyses of his patients, he concluded that stories about earth spirits, sky gods, and all the other creatures that make up the pantheon of the supernatural are ways of describing the behavior of neural circuits within the unconscious. Our inheritance of myths, fairy tales, and the magical arts provide us with a map of this hidden world. Stories about turning lead into gold, befriending wild animals, and transforming cruel monsters into gentle princes are stories about the unconscious. These stories teach us how to facilitate cooperation between the two parts of the mind to achieve an integrated self in which the two work in harmony.

Since the unconscious is, by definition, outside of conscious awareness, it can't be observed directly. In some ways, it's like the dark matter that, physicists hypothesize, holds galaxies together. Dark matter is thought to account for 85 percent of the matter in the universe, and it exerts a powerful influence on the universe's structure, but it can't be detected by any instrument we've yet developed. We only know it exists from observing its effects on things we *can* measure.

The unconscious is the same. Although we can't see it, it profoundly influences our lives. Usually it stays in the background, going about its work unnoticed. Occasionally, however, unconscious contents erupt into consciousness, and a sort of possession occurs, often in the form of passion, enthusiasm, inspiration, or a newfound strength we didn't know we had—and the effects of that possession, we *can* observe.

The first section of this book will examine those effects in daily life. It will also review what science has discovered about how the unconscious works. (This part may seem far removed from magic, but it's essential to understand the ordinary influences of the unconscious before we dive into its mysterious depths.) We'll also see how Jung's understanding of the unconscious leads us from scientific experiments to mystical experiences.

The second section will explore ancient stories of mysticism and several important magical traditions. Stories are more important than most people think. Human brains need them to make sense of the world. Things can't just happen; they must happen for a reason. Our minds shape events into stories in order to give them meaning. We're generally unaware of it, though, because the narratives are constructed within the unconscious, which makes them subject to the influence of psychological dark matter. For psychologists, stories are like the galactic phenomena that astrophysicists study to learn about dark matter; they reveal what can't be seen.

The best stories for this purpose are the oldest ones. Most stories are of little interest a generation or two after they're told, but some last for thousands of years. They do so because they effectively embody the dynamics of our inner world. Not surprisingly, the oldest stories we still tell today, legends and myths, are stories about the supernatural. The themes in these stories—mythological motifs, as they're called—are sometimes borrowed by modern writers to give their work narrative muscle. We'll look at some of these modern examples to illustrate ways in which those ancient ideas still influence our culture today. But primarily, the stories that will open a window for us into the unconscious in these pages are fairy tales, alchemical speculations, and philosophical musings about the mystical properties of numbers. We'll also explore the symbolic illustrations of the mystical tarot deck, which vividly depict the unconscious forces that shape our destinies from within.

Each chapter in the second section highlights the ultimate goal of Jungian psychology: the integration of the conscious mind with the unconscious—the unification of all aspects of the psyche to form the complete self, one's true identity. The final section of the book will detail this process further and examine how it leads to transcendence, a state of balance and contentment arising from the ability to go beyond a focus on one's personal needs and attain a more cosmic perspective. We'll explore both scientific and mystical ways of understanding transcendence and then look at an ancient discipline that provides one path to this unification. Transcendence may be associated with sages and mystics, but it can also be achieved by ordinary people, if they are willing to undertake the difficult climb that leads to life's highest summits. This final section will show how it's done.

Throughout the book, we'll examine what it feels like when our behavior is coopted by the irrational, uncontrollable forces of the unconscious, and the ways this metaphorical possession can induce non-ordinary states of consciousness. These states may be subtle or overwhelming, but they always take us beyond the experiences of daily life. Whether they inspire, terrify, or transform, they always make us feel more alive. Meddling in the world of the unconscious is dangerous, as it can unleash a torrent of uncontrollable passions, irrational convictions, and destructive impulses that can upend a person's life. But the rewards of tapping into this primordial source of energy, life, and inborn wisdom justify the risks.

It's not easy to accept that a large part of one's mind is under the influence of forces that lie hidden, beyond our control. Maybe we can't always control our bodies, but surely we ought to be able to control our own minds.

So we imagine ourselves sitting alone inside our heads, just behind the eyes, never turning around to look into the darkness behind us—a darkness teeming with magical creatures who help us and hurt us every day.

This is a book for the adventurous—for those who have the courage to turn around and look.

Part I

THE UNCONSCIOUS

INTO THE DARKNESS

We believe that we are masters in our own house only because we like to flatter ourselves. In reality we are dependent to a startling degree on the proper functioning of the unconscious psyche, and must trust that it does not fail us.

C. G. JUNG

THE EGO AND THE UNCONSCIOUS

That annoying song was going through Jane's head again. She hadn't heard it in years. Now she couldn't stop humming it. As she made her way down the sidewalk toward her office, a woman coming in the opposite direction smiled at her, and her mood lifted. She walked on, and as she passed a coffee shop, a pang of hunger reminded her that she needed to get her blood drawn. Her doctor had told her not to eat or drink for twelve hours prior to the blood draw, so she had skipped breakfast. It was her second attempt this week. Yesterday she had gotten halfway through a muffin before remembering the blood test. Even today she felt tempted to stop at the coffee shop for a bagel.

As she kept walking, she noticed a growing feeling of anxiety. She searched her memory for a possible cause and remembered she had a meeting with an angry client later that morning. It wasn't her fault the order had gone out late, but she was the one who had to make the apology. A moment later the worried expression on her face cleared as she thought of a good way to smooth things over.

The song, the lift in mood, the hunger, the blood test memory, the temptation to eat, the anxiety over the meeting, and the solution to her difficulties—these reactions to her internal and external environment all took place within Jane's mind, but they came unbidden. That is, she didn't choose to experience any of them. Moods, memories, and inspiration are all outside our conscious control. They're visitors who drop by unexpectedly, sometimes welcome, sometimes not. But if she didn't summon them, who did?

Finding the answer begins with a closer look at the part of the mind we're familiar with, the part we refer to when we use the word *I*.

The Latin word for *I* is *ego*. In English, the word *ego* has a negative connotation because it's typically associated with being conceited. But in Latin, it's neutral. It simply translates to the word *I*. Psychiatrists use the word *ego* as a technical term for the person *I* refers to, as in, "I enjoyed that movie." They use this technical term because the ego, the I, is not the only personality inside your head. There are many others, and although you may not be aware of them, they control your life in profound ways.

The ego is described as conscious, because it's the part of your mind you're aware of. Whereas the ego is conscious, the rest of the mind is unconscious, because you generally don't experience it directly. Because of this, when the unconscious interacts with your ego, it can feel like an external force. Your mental experience doesn't seem to be originating within the "I" you identify with.

The ego has less processing power than the unconscious. It can do only one thing at a time. The unconscious, by contrast, contains circuits that work in parallel, running many processes at once, so it's capable of more sophisticated tasks. Even as Jane's unconscious was passing up to her ego all the things she experienced on her walk, it was busy with a multitude of other tasks. It was coordinating millions of individual muscle fibers needed to maintain her balance and propel her down the sidewalk. It was controlling her heart rate and breathing to ensure adequate oxygen flow. It was orchestrating the secretion of hormones into her bloodstream and processing signals arising from the interaction between photons in her environment and receptors in her eyes, turning spikes of electrochemical activity into vision.

Depending on what it's doing, your ego can process between 10 and 60 bits of information per second.[1] To put that in perspective, reading this sentence requires you to process about 45 bits per second. That doesn't leave much room for anything else. It's unlikely that you can continue to read this page while simultaneously planning what you'll make for dinner. But this limitation applies only to your conscious mind. Your brain as a whole has a much greater capacity for mental activity, managing more than 11 million bits per second. When it comes to processing information, the unconscious is half a million times more powerful than the ego. Consciousness is a tiny boat floating on a vast sea of darkness, a sea that's teeming with life, pushing the boat from beneath the surface.

What role did Jane's conscious mind play as she walked to work? Besides being aware of what was going on, the most significant thing her ego did was decide to resist the urge to walk into a coffee shop and order a bagel. The conscious mind, the person you call "I," plays a smaller role in your mental life than you may realize, but it's an important one. One of the things it does is evaluate instinctual impulses that emerge from the unconscious and decide whether to act on them.

It's not easy to accept the idea that we're not alone inside our heads. When Sigmund Freud, an early pioneer in the study of the unconscious, claimed that most mental activity takes place outside of conscious awareness, he met strong resistance.[2] His contemporaries were convinced that mental activity and consciousness were one and the same. The existence of thought without awareness was too hard for them to accept. However, a large body of research has since confirmed the existence of the unconscious by measuring its effects and how it influences our behavior in pervasive, sometimes dramatic, ways.

You may not realize how much you rely on your unconscious. Managing bodily housekeeping functions is only the beginning. It also plays an important role in processing information. If I were to ask you how many people live in New York City, you would probably do an internet search to get the answer. You would be consciously aware of each of the steps you took, from opening a browser to typing the query. By contrast, if I asked you to tell me your mother's maiden name, you'd come up with the answer effortlessly, but you wouldn't be able to tell me how you found it. The answer was simply presented to you.

The same thing happens when you speak. Consciously, you know what you want to say, but you have to rely on your unconscious to supply the words. Most of the time your unconscious cooperates, but not always—like when the word you're looking for is on the tip of your tongue. The ego knows what it wants to say, but the unconscious won't deliver the word.

Sometimes the unconscious delivers a word the ego doesn't like. It may offer something insensitive or that isn't quite right. At those times the ego can exercise its veto power and try to find something different. Occasionally, the unconscious delivers something quite different from what the ego intended, and the ego fails to notice the switch. That's called a "Freudian slip," and it reveals the unconscious has intentions of its own.

There was a college professor who wasn't well liked by the women in his class. They felt he favored the men, an accusation he vehemently denied. Things boiled over when, as part of an exam's instructions, he wrote "A pen is required for this test." At least that's what he intended to write. Unfortunately for him, the accidental omission of a space between the second and third words appeared to confirm the students' suspicions.

It's not unusual for the unconscious to get us into trouble. Jung writes that unconscious contents may "behave like Descartes' devils and seem to delight in playing impish tricks. They slip just the wrong word into one's mouth, they make one forget the name of the person one is about to introduce, they cause a tickle in the throat just when the softest passage is being played on the piano at a concert, they make the tiptoeing latecomer trip over a chair with a resounding crash."[3]

The unconscious can be treacherous, but it can also serve as a friendly helper. Whether we're giving a presentation, discussing ideas with a coworker, or socializing at a party, there are times when we're surprised by our own eloquence. We feel inspired, and the rugged path between our thoughts and our words has, for a little while, been made smooth.

The unconscious exerts a surprising amount of control over our behavior. You can consciously choose to step on a treadmill, but can you maintain a regular exercise routine? It can generate new ideas and emotional states. You can update a spreadsheet, but can you summon a brilliant idea? You can plan a pleasurable activity, but can you be sure you will enjoy it? The unconscious mind is responsible for the things we value most in life: love, joy, friendship, inspiration, compassion. Conscious planning can make these things more likely to occur, but ultimately, we have to trust and hope that our unconscious comes through for us. Sometimes the unconscious cooperates and delivers what the ego wants; at other times it has plans of its own.

When you go to a party, you can't control whether you will be engaging and fun. Sometimes it happens and sometimes it doesn't. When you're with a loved one, you can't control whether you will experience an emotional connection. Sometimes you can't even control whether you will be kind and thoughtful or snappish, irritable, and cruel. Not everything that comes from the unconscious is good. In addition to love and joy, it also produces hatred, depravity, and evil.

But it's a wonderful feeling when the goals of the unconscious are aligned with those of the ego. Think about the difference between working on a project that excites you compared to one you're doing simply because your ego decides it must be done. When you're fascinated by your work, your unconscious is delivering energy, emotion, and other mental resources. Time flies by. Each day you look forward to getting started, and the work brings out the best in you. On the other hand, when you're doing something just because you have to, you lack unconscious support. All you have is the best efforts of the ego. Whether a homework assignment, task at work, or household chore, it's exhausting, because only the unconscious can supply mental energy. The ego has very little of its own. Without the unconscious powering the task, you must force yourself to keep at it, and you'll look for any distraction. Email, social media, even cleaning the kitchen cupboards is preferable to writing the damn report. There's a big difference between those two states of mind. Passion doesn't obey the ego's call. But it's of essential importance. It determines whether you'll be happy or miserable as you work, and whether the results will be outstanding or merely adequate.

The unconscious is so powerful, and so different from conscious thought, that when it's activated, it can be experienced as a kind of possession, a visitation from a foreign entity that can feel friendly, hostile, or a little of both. When the unconscious makes its presence felt, Jungian psychologists call it an *eruption*, *invasion*, or *incursion* into consciousness. Sometimes these eruptions are destructive, producing overwhelming emotions that cloud rational judgment, destructive impulses that may ruin a person's life, and desires that run counter to the values of the conscious mind—and are accompanied by an irresistible pressure to fulfill them.

Other, more positive incursions can come in the form of intuition, valuable insight we experience as a gut feeling, or as inspiration. An idea appears out of nowhere, carrying us far beyond what conscious reasoning is capable of. It sometimes feels like a gift from the gods. An inspired idea may solve an intractable problem, create a work of priceless beauty, or even change the course of our life.

Eruptions from the unconscious can also give people abilities they don't have under ordinary conditions. Enthusiasm enhances a person's capacity for work. Panic may give rise to superhuman strength, and just being "on" can make an athlete perform feats of agility well beyond her usual skills. Once in a while, the divine possession of an unconscious eruption reveals to us what seems to be a deeper reality. It may be a magic moment, a mystical encounter, or enlightenment. These experiences can take us to the distant heights of human existence. But before we explore these extraordinary states, we need to take a closer look at the more ordinary influences of unconscious processing.

FRAGMENTARY PERSONALITIES AND DIVINE POSSESSION

Different areas of the brain are specialized for different functions. The part of the brain that lies just behind the forehead is called the prefrontal cortex; it's larger in humans than in any other animal and is a recent evolutionary development. It plays an important role in planning long-term goals, making decisions, and keeping social behavior within acceptable limits, all important functions of the ego. The medulla, which is located where the brain connects to the spinal cord, is responsible for simpler functions, such as controlling breathing, heart rate, and blood pressure. The medulla is more concerned with the functioning of your body than your psychological experience.

Circuits in the brain act relatively autonomously. They can cooperate with one another or they can compete. Competing circuits are common. Circuits that process vision, for example, compete in such a way that if you're paying attention to one thing, let's say a grilled cheese sandwich, the other elements in your visual field, such as the design on the plate, are less noticeable. That way, you aren't distracted from what's important. If you were a chinaware collector, on the other hand, the circuits focusing on the dish design might win the competition, and the sandwich would fade into the background.

As Jane walked to work, we saw the competition between the part of her brain that wanted to eat breakfast and the part that wanted to follow her doctor's orders to get a fasting blood test. On the previous day, the muffin-craving part had won by making the conscious circuits forget about the blood draw. The second day, the conscious circuits won and overcame the craving for a bagel.

The human mind is a seething cauldron of these competing and cooperating drives. Psychologist Michael Gazzaniga called it an orchestra without a conductor.[4] The ancient Greek theologian Clement of Alexandria compared man to the mythical Trojan horse, "embracing as he does in one body a host of such different spirits."[5] Depending on one's circumstances, different behaviors, even different personalities, come to the surface. Your boss, who you've never seen smile, might turn into a comedian at home, but only when she's horsing around with her three-year-old son. That side of her personality can't express itself in other environments. It's not something she can choose to put on or take off at will.

Similarly, you may have a group of friends who bring out a side of you that is seen in no other circumstances. When you get together, you may act uncharacteristically serious or bold or utterly immature. When you lose a friend, you not only lose that person, you also lose the part of yourself that person brought out. When a friend dies, part of you dies as well.

We all have diverse personality pieces within us. They're not complete personalities, as we see in sensationalistic accounts of people with "multiple personality disorder," a rare, abnormal condition. But everyone experiences mysterious changes in their behavior from day to day, even moment to moment. Sometimes we're confident; other times we're shy. Sometimes we take silly risks; other times we play it too safe. Sometimes these unconscious personality pieces are sophisticated enough to develop their own goals and the agency needed to bring those goals to fruition. When they occur within the brain, these agents are sometimes referred to as "fragmentary personalities." They exhibit perception, feeling, and intention.

Drug addiction is an illness that provides us with a uniquely clear view of a fragmentary personality at work. Alcoholism develops as a result of the interplay among alcohol use, the environment, and genetic vulnerability. This interplay can give rise to a personality fragment that values alcohol over everything else—a "thirsty demon," as it's been called. One strategy the thirsty demon uses to get its way is the production of craving. We've all experienced craving, whether for a piece of cake when we're on a diet or a little more time under the warm comforter when the alarm clock goes off on a chilly winter morning. Craving diminishes the ego's ability to make voluntary choices. You consciously intend to stick to your diet, but your ego isn't the only agent involved in the decision. Sometimes a personality fragment gets its way with an addict by simply seizing control. Many smokers have found themselves puffing on a cigarette with no memory of taking it out of the pack.

It's not only in the context of addictions that we see fragmentary personalities. Your spouse might say, "What's wrong with you today? You criticize everything I say." You might be unaware you're behaving any differently, but it's probably not with your usual urbanity. Fragmentary personalities act in more primitive ways compared to the conscious ego, and when they become too strong for the ego to handle, we can get into trouble. One notorious example is the way sexual arousal can affect people's behavior and the choices they make.

Researchers in the University of Windsor Psychology Department were interested in examining the relationship between sexual arousal and risk taking.[6] They showed some volunteers sexually explicit video clips and others neutral ones, and then had them fill out questionnaires about how they would act in hypothetical romantic scenarios. The men and women who experienced arousal both reported greater intentions to participate in risky sexual behavior, such as having unprotected sex with a new partner. This personality fragment was willing to put the volunteers in harm's way to get what it wanted.

The volunteers shown sexually explicit video clips also took more risks outside the narrow realm of sexuality. When the study subjects later played a computerized game of blackjack, the ones who were sexually aroused showed a greater willingness to make risky plays, such as asking for a card when they were close to twenty-one. When the contents of the unconscious are activated, in this case by stimulating the reproduction instinct, behavior is no longer directed by the rational ego alone. Agents that exist outside the sphere of conscious control compete with the ego, and often win.

We're all familiar with examples of smart people doing foolish things—making shockingly unwise financial decisions or indulging an impulse that destroys their career. As we noted earlier, a typical response to this kind of behavior is to ask, "What on earth possessed her to do that?" or "What could have gotten into him?" These common expressions reflect the half-acknowledged belief that we are liable to being taken over by forces beyond our control.

The unconscious can steer us toward trouble, but it can also show us the buried treasure locked inside ourselves. The German philosopher Arthur Schopenhauer writes, "I have familiarized myself with the factual data of a theoretical and practical problem; I do not think about it again, yet often a few days later the answer to the problem will come into my mind entirely from its own accord; the operation which produced it, however, remains as much a mystery to me as that of an adding machine."[7] The mathematician Henri Poincaré had a similar experience: "At that moment when I put my foot on the step the idea came to me, without anything in my former thoughts seeming to have paved the way for it, that the transformations I had used to define the Fuchsian functions were identical with non-Euclidean geometry."[8]

Creative workers often describe themselves as bystanders[9] to whom new ideas or solutions to problems are revealed seemingly out of nowhere. Many speak of having a muse or receiving divine inspiration. The "I" that speaks through the artist's mouth cannot take the credit.

We think of inspiration as a flash of insight happening in a moment. A new idea is born, and then the moment is gone. But when an idea is inspired, the unconscious provides not only the idea but also the energy and motivation needed to make it a reality. Inspired inventors, artists, and writers are notorious for working at all hours of the day and night. They skip meals and forgo showers and clean clothes for days at a time because they're so engrossed in their work. They're filled with enthusiasm, a word derived from the Greek *entheos*, meaning "possessed by a god." This time the possession is not by a demon bent on destroying a person's life; it's possession by a divinity giving birth to creation.

Whether it's an addict possessed by his thirsty demon or an artist inspired by her creative muse, the actions of the unconscious connect everyday life to stories about the supernatural. These stories are about powerful entities who are mysterious and cloaked in darkness, like the neural circuits within the unconscious. They're morally ambiguous. Sometimes they help humans, often in astounding ways, but they can be capricious, and will harm as readily as they help. They don't follow the rational rules of conscious thought, but their behavior isn't random. They have rules of their own that can be learned by those who are (un)fortunate enough to stumble into their realm.

THE MAGICAL PERSPECTIVE

Seeing the world through the eyes of our hidden fellow travelers is an essential step in attaining our full human potential, but it is not for the faint of heart. The irrational nature of the unconscious can frighten the ego, and appropriately so, given the ability of the unconscious to overwhelm the ego, robbing it of its ability to think clearly and limit its instinctual impulses. The ego has to manage its relationship with the unconscious carefully. If the ego pulls back too far, it becomes weak and unable to function effectively, a bloodless calculating machine cut off from its instinctual roots. If it gets too close, it may sink into the primordial darkness of the unconscious, descending to the level of its animal nature. We see this happen during states of intoxication, when the prefrontal cortex is so impaired that the ego is unable to function. Under these conditions, people may give free rein to the more primitive aspects of their personality, such as rage, lust, or maudlin affection. Although the unconscious can be a powerful ally, we must proceed with caution. Uncontrolled passions have destroyed many lives and even entire nations.

While it's prudent to be afraid, we shouldn't reject this essential part of our humanity. Beginning with the Enlightenment, Western culture has become infatuated with the rational world of the ego, particularly with science, the greatest accomplishment of the conscious mind. The scientific endeavor has been so successful that many people believe we will master the world through calculation and achieve an earthly paradise in which all our needs are met. But humans have needs that go beyond the material, and science doesn't help with the development of our inner world. It only has tools to control our environment—the outer world. If we don't recognize its limitations, science may tempt us to seek a life of safe, controlled predictability and deceive us into thinking we're following a path that will lead to happiness.

Unfortunately, happiness doesn't work that way. There's little correlation between it and material wealth (at least after one's basic needs are met).[10] Although we may become obsessed with material goods, they won't satisfy us. For that we need immaterial things like love, friendship, beauty, creativity, and growth. Science doesn't help us with these things, but our inheritance of art, myth, music, and poetry—works inspired by the unconscious—can help a great deal. Suppressing the magical instinct that comes from the unconscious may appear sensible in this age of scientific rationalism. But it cuts us off from an important part of our humanity. It's like seeing the world in black and white, when it could be filled with color.

Let's look at an example. What is air? Scientists tell us it's primarily nitrogen, oxygen, and carbon dioxide. If you understand that, it will help you understand the growth of plants and many other useful things. But as one mystical author writes, "What are its oxygen, its hydrogen, its nitrogen, its carbonic acid, its ozone, and all the possible rest, to the blowing of the wind on our faces? What is the analysis of water to the babble of a running stream?"[11]

Psychologist William James writes, "Weight, movement, velocity, direction, position, what thin, pallid, uninteresting ideas! . . . It is the terror and beauty of phenomena, the 'promise' of the dawn and of the rainbow, the 'voice' of the thunder, the 'gentleness' of the summer rain, the 'sublimity' of the stars, and not the physical laws which these things follow, by which the religious mind still continues to be most impressed."[12] James was writing specifically about the psychology of religion, but we could easily replace "the religious mind" with "the poetic mind" or even "the human mind."

Walt Whitman expressed a sentiment similar to James's in his poem "When I Heard the Learn'd Astronomer":

When I heard the learn'd astronomer,
When the proofs, the figures, were ranged in columns before me,
When I was shown the charts and diagrams, to add, divide, and measure them,

When I sitting heard the astronomer where he lectured with much applause in the
lecture-room,
How soon unaccountable I became tired and sick,
Till rising and gliding out I wander'd off by myself,
In the mystical moist night-air, and from time to time,
Look'd up in perfect silence at the stars.

Whitman contrasts the use of analysis, which is the hallmark of conscious thought, with the way the unconscious understands the world. *Analysis* is derived from a Greek word that means "to cut into pieces." A chemist understands air by dividing it into its component elements. A botanist understands a flower by separating the roots, stem, and petals. At the end of Whitman's poem, the analyzing ego is silenced, and the narrator engages in what might be described as passive appreciation. He obtains a kind of intuitive understanding that can't be expressed in words.

Is an intuitive, mystical understanding also a useful one? If it is, it must be useful in a different way than we're accustomed to imagining. Science makes our lives easier. It helps us get the most work done with the least effort. From the internal combustion engine to artificial intelligence, science makes machines do more, so we can do less. Magic is the opposite. Myths, fairy tales, and legends are stories about struggle. The lazy and dishonest don't get far. But those who are diligent and self-sacrificing overcome the challenges they face and, in the end, achieve their goals. Magical stories are about becoming the person we were meant to be through hard work and sacrifice. Just as a weightlifter needs resistance to build muscles, humans require obstacles in order to grow.

Magic is what we live for. We think we want material goods—a larger TV, designer clothes, the latest cell phone—but what we really want is to feel alive. An expensively furnished room is nice to look at, but the wind blowing through your hair on a cool autumn morning transforms your entire world. It's what we call a "magic moment." It may not be strictly supernatural, but we use the word *magic* to describe it because something extraordinary happens that's different from our ordinary state of consciousness.

Physicist Alan Lightman writes about a magic moment he experienced watching two baby ospreys take flight for the first time at his home in Maine:

> When they were within twenty feet of me, they suddenly veered upward and away. But before that dazzling and frightening vertical climb, for about half a second we made eye contact. Words cannot convey what was exchanged between us in that instant. It was a look of connectedness, of mutual respect, of recognition that we shared the same land. After they were gone, I found that I was shaking, and in tears. To this day, I do not understand what happened in that half second. But it was one of the most profound moments of my life.[13]

A magic moment wakes us up from our familiar routines. It makes us realize the world has a deeper, more mystical dimension than we're accustomed to seeing. They occur unexpectedly, sometimes triggered by seemingly ordinary events: a shared moment of intimacy, a sunset, or the smell of autumn leaves. We have no control over when they occur, but when they do, we connect with the unconscious, the ordinary fades away, and we remember what it feels like to be fully alive.

THE SOCIAL INSTINCT

Magic moments draw our attention away from ourselves, inviting us to look outward. The boundary that separates the personal from the universal becomes more fluid, sometimes disappearing completely. When the unconscious is activated, it tends to promote unity and integration. It complements the ego's focus on analysis—the separation of one thing from another—and helps us connect with other people and the world at large.

Like all products of natural selection, the human brain evolved to keep members of our species alive and reproducing. These are selfish goals, and it makes sense that natural selection would lead to behaviors that prioritize one's own survival over that of others. People who work to maximize their own resources, such as food, shelter, tools, and reproductive partners, are more likely to pass along their genes to the next generation, which further enriches the pool with genes for selfish actions. After thousands of generations, we might expect selfishness to have become the guiding principle of human behavior.

But that's not what we see. Humans frequently act in unselfish ways, making sacrifices for the good of others. Although that may seem counterintuitive from an evolutionary perspective, when humans work together as a group, they can be more successful than individuals acting alone. No matter how big or strong an individual is, he won't hunt, build, or fight as well as a smoothly functioning team. Consequently, evolution favors organisms that cooperate. Once again, we're faced with a pair of competing principles that need to find balance: in this case, between selfishness that maximizes the success of an individual's genes and cooperation that maximizes the success of the species.

The conscious mind tends to represent the needs of the individual. This fact may help explain how *ego*, the Latin word for I, became associated with being self-centered. Some philosophers go so far as to argue that the conscious mind is incapable of engaging in genuine altruism. It's a theory called "psychological egoism," and it holds that everything we do is ultimately motivated by our own self-interest. What appears to be altruism is really a desire to make ourselves feel good, polish our reputation, or acquire some other personal benefit.

Imagine a friend needs money and asks you for a loan. It's more than you can conveniently afford, and you're pretty sure you're not going to get your money back, so you might decline because of the expected financial loss. On the other hand, you might give them the money because you want a reputation for generosity or to maintain the friendship. The conscious mind bases decisions on reason, and reason weighs the pros and cons of each alternative. This weighing of options is inherently selfish because the underlying question is always, "What will provide me with the largest benefit?" It's possible the benefit could be an emotional one. You might decide to give because it feels good, or alternatively, because acting selfishly would trigger feelings of guilt. You might also believe that if you adhere to a higher moral code, you'll receive your reward in the afterlife. There's always a calculation.

The theory of psychological egoism may insist that pure altruism is beyond the capabilities of the ego, but that doesn't mean it's beyond the capabilities of a person. Your friend's request for a loan could trigger a generous impulse, leading you to acquiesce without really thinking about it. That would be your unconscious making the decision for you, and it doesn't involve a calculation of relative benefits (as far as we know). Such an unconscious impulse is more of a biological drive to be generous, much like the drive to eat when you're hungry.

In addition to producing generous impulses, the unconscious also promotes altruism via emotional reactions to the choices we make. We tend to act in ways that make us feel good, and we avoid doing things that cause us emotional pain. Those feelings are generated by the unconscious, which opposes selfish behavior through emotional reward and punishment. If you give your friend the loan to avoid feelings of guilt or because helping them feels good, your ego is acting out of self-interest, but your unconscious is not.

It's odd to have this force of justice—this entity so powerful, it can make us act contrary to our own self-interest—manipulating us from out of the darkness. In classical times, such impulses were attributed to the gods, just like passionate love and martial courage. An ancient Roman might believe a generous impulse came from the goddess Liberalitas. In the Judeo-Christian tradition, the Bible quotes God as saying of the Israelites, "I will put my law within them, and I will write it on their hearts."

The tendency of the unconscious to promote unity can be seen in the two fundamental tenets of mysticism. The first is "all is One." The revelation of the essential unity of all things is central to the mystical encounter, whether religious, spiritual, or drug induced. The other mystical concept is universal love, the highest practice in many religious and spiritual traditions. It's the realization that all things—people, animals, even plants and inanimate objects—deserve our compassion. As we'll see in the next chapter, the ego may view its environment as a collection of practical resources that can be put to good use. But to the unconscious, it's a world filled with friends.

SPIRITS EVERYWHERE

Everything that is dead quivers. Not only the things of poetry, stars, moon, wood, flowers, but even a white trouser button glittering out of a puddle in the street . . . Everything has a secret soul, which is silent more often than it speaks.

WASSILY KANDINSKY

"I TRY TO LISTEN TO THE ROCKS"

Lucas Congdon is America's most famous swimming pool builder. His company, Lucas Lagoons, builds some of the most creatively unique pools and natural rock lagoons in the industry. Their creations have earned them more than fifty design awards, and in 2015 the Animal Planet network debuted a reality TV series featuring Congdon's company.

Congdon was born into a family of Vermont stonemasons. He began cutting and crafting stone when he was fourteen years old, and stones are what have made him famous. In one episode of the television series, Congdon explains that the process of picking out the rocks he uses for his projects begins with a walk in the woods to look at natural formations. "I look for rocks with channels and moss and lichen," he says. "Then I try to listen to the rocks." He looks at a rock and softly says, "Talk to me. What do you want to be, silly little rock? People may think I'm a little nuts because I talk to rocks." He hoses down a six-foot slab, and says to it, "Really see the water hitting you on the way down, working through your channels." He turns back to the camera, and says apologetically, "But you've got to be a little crazy to work with rocks all day."

Talking to rocks may sound crazy, but when it's being done by an award-winning designer, we shouldn't write it off too quickly. Is it just an eccentric habit Congdon picked up, or might it play a role in his success? Evidence suggests that unconscious circuits in Congdon's brain help him build better pools by inspiring him to see the world from a magical perspective, one in which rocks are alive and can talk to him. He does better work when he treats his rocks as friends rather than simply as construction materials.

SCIENCE AND RELIGION

The circuits that encourage Congdon to listen to rocks—those responsible for the experience of empathy—evolved for the practical purpose of facilitating cooperation with other people. But when these circuits create a feeling of connection not only with other people, but also with spirits who seem to inhabit the things around us, they take us beyond the practical concerns of daily life and lead us to the supernatural. The term *supernatural* refers to anything outside nature, which is understood as consisting of matter (fermions, in scientific terms), force (bosons), and nothing else. A view of the world that denies the existence of anything beyond fermions and bosons is called "materialist." Modern Westerners are quasi-materialists. Most believe in the existence of human souls and probably a deity, but that's all. Everything else is fermions and bosons.

Many scientists endorse strict materialism, denying the existence even of human souls. Francis Crick, who, with James Watson, discovered the double helix structure of the DNA molecule, described the strict materialist perspective, which he named the Astonishing Hypothesis, in the following terms:

> The Astonishing Hypothesis is that "You," your joys and your sorrows, your memories and your ambitions, your sense of identity and

> free will, are in fact no more than the behaviour of a vast assembly of nerve cells and their associated molecules. As Lewis Carroll's Alice might have phrased it: "You're nothing but a pack of neurons."[1]

It's "astonishing" because it contradicts our intuition about who we are. It's also provocative because, for many people, spiritual beliefs are the most important beliefs they hold. A book such as this one, which explores the role of supernatural phenomena in mental life, will unavoidably touch upon spirituality and religion. It's a sensitive topic, so it's important to be clear about the limitations of science.

Explaining something (for example, describing what happens in the brain when people engage in spiritual practices) and explaining something *away* are not the same thing. Science and theology are independent fields of study. Scientists study matter and force; theologians study spirit. The two disciplines don't intersect because they concern themselves with different subjects. Spirits, if such things exist, are immaterial, containing neither fermions nor bosons, and although science can give an account of spiritual phenomena in material terms—for instance, which parts of the brain are active when someone has a religious experience—science can't provide any evidence, either for or against, supernatural causes of those phenomena.

It's like looking at the brain's response to an encounter with a lion. We would probably see increased activity in the amygdala, corresponding with alarm, but that same response could be the result of watching a video of a lion or hearing a vivid description of a lion as easily as seeing a real lion. Similarly, if there are structures in the brain that are active when people report an encounter with God, science has no way of determining whether those structures were activated by an actual deity or by something within the material realm.

Although we can give scientific explanations for mystical experiences and explore why it's important to take those experiences seriously, we can't say anything about the causes of mystical phenomena, because that's a question of faith, which falls outside the scope of what science can study. Along the same the lines, this book will use the Bible to illustrate some of the ideas we'll encounter. Just as neuroscientists study the brain without addressing the question of soul, we'll look at the Bible as the most important piece of

Western literature dealing with the supernatural, but without addressing the nonscientific question of whether it's based on divine revelation.

Materialism may or may not be the most accurate description of objective reality, but it's not an accurate description of subjective experience—and that subjective experience, for human beings, is one of a world filled with spirits.

ANIMISM

We may not always notice it, but we, like Congdon, naturally fall into the habit of treating things around us as if they were alive. The technical term is *animism*, derived from *anima*, the Latin word for spirit. It's the belief that objects, such as rocks, rivers, forests, oceans, even artificial things, are sentient. Psychological research can help us understand how animism enhances our ability to interact with the world effectively—in other words, why listening to rocks helps Congdon build better swimming pools. But before we look at the explanation, let's examine the phenomenon itself and get a sense of just how pervasive it is.

We have many stories about objects that are alive, but they're mainly written for children. These objects include the sorting hat in the Harry Potter series, the fighting apple trees in *The Wizard of Oz*, and the brooms that torment Mickey Mouse in *The Sorcerer's Apprentice*. Some of the earliest cartoons were primarily about everyday objects coming to life. Perhaps it's not surprising that this new art form of animation was used to illustrate animistic fantasies.

The Academy Award–winning movie *Who Framed Roger Rabbit*, which celebrates the early years of animation, contains countless examples of ordinary objects coming to life. The scene in which detective Eddie Valiant makes his return to Toontown has talking birds, trees, mushrooms, rocks, flowers, butterflies, mice, rabbits, and classical deities, as well as living cars, buildings, clocks, suitcases, and furniture.

Children perceive the world through the lens of magic because their conscious minds are only beginning to develop, meaning they're primarily influenced by unconscious processes. For example, young children believe

Early animation was used to bring inanimate objects to life.
(Source: The Big Cartoon Database.)

that wishing for something makes it come true. If a toddler is angry at her brother, and he becomes sick, she may think the illness is her fault. Children first begin using reason around age seven, and their facility with it gradually increases through adolescence. As that development occurs, their magical way of understanding the world fades away, to be replaced by a more scientific model that does away with spirits. Animism is replaced by materialism.

Eugene Vasilevitch Subbotsky, a psychologist at the University of Lancaster, England, wondered what happens to the magical point of view as we get older. He thought that perhaps it's not so easy to rid oneself of the magical perspective, and that adults were able to adopt a materialist perspective only by building up defenses against the magical view. Subbotsky wanted to see if he could break through those defenses, bringing the old magical beliefs back to the surface. He did it using a strategy he called "raising the stakes."[2]

Subbotsky recruited a group of British adults who said they didn't believe in magic and told them a story about a witch who approached a man walking down an empty street. The witch offered to put a spell on him that would make him rich and happy. Subbotsky asked his research participants whether they would advise the man to accept the spell or decline it.

Half the participants suggested he should accept it, arguing that he should do it to please the witch. The other half said he should decline the spell to avoid encouraging the witch. When the story was changed, so that the spell became a curse that would make the man a servant of evil forces, Subbotsky got the same results. Half the participants said they would advise the man to accept the spell and half advised he decline it.

Then Subbotsky raised the stakes. He said to the research participants, what if the witch approached *you* and offered a spell that would make *you* rich and happy? Would you accept or decline the spell? Now that it was personal, the numbers changed. Sixty percent of the research volunteers said they would accept the good spell. When they were faced with the bad spell that would make them a servant of evil forces, every one of them declined. Not a single person was willing to accept this spell, despite a professed disbelief in magic. Theory is all well and good, but when it comes to practical matters, matters that might impact one's well-being, people take magic more seriously.

The research volunteers didn't know they believed in magic. They thought they were proper, modern materialists, and they probably weren't impressed by the results. After all, stories about witches don't have much to do with everyday life. But Subbotsky disagrees. He writes about playing games of imagination with his young son, who had not yet outgrown the magical perspective. "What impressed me most in my son's attitude . . . was his absolute confidence in the idea that nature would be kind to him; that the elements (water, air, gravity) were aware of his presence in the universe, accepting his divine right to be there."[3]

As we grow into adults and lose this belief, the universe becomes a cold place, governed by the impersonal laws of physics. "Of course," writes Subbotsky, "we are in debt to science for its remarkable achievements—modern medicine, elaborate technologies, relatively comfortable lives. Yet the price of such comfort is great: The loss of a meaningful connection between man and nature."

When adults express animistic beliefs—for example, "This machine hates me"—they claim to be speaking figuratively. But experiments like Subbotsky's suggest they may not fully understand their own minds. You may be able to identify animistic tendencies in yourself by imagining how you would feel if you checked into a hotel and the desk clerk put you in a room where someone had been murdered. A true materialist would fail to see any connection between an event that happened in the past and getting a good night's sleep tonight. But most of us would feel that something disquieting lingered within the room.

A real-life example involved a young couple who had recently moved into a new home. They had been house-hunting for months but couldn't afford any of the ones they liked. At last, they found a wonderful house in a desirable neighborhood that was within their price range. It was only after moving in that they learned from their neighbors the previous owner had been murdered in the upstairs bathroom. They were so upset, they thought about suing the seller or putting the house back on the market.

Fortunately, the story has a happy ending. Although they weren't particularly religious, when a friend suggested they ask a priest to bless the house, they agreed. The priest came over and prayed for the murdered man's soul to be at peace. He went from room to room swinging an incense burner, purifying the house with its fragrant smoke. When it was over, the new owners were once again thrilled with the bargain they had found.

These kinds of experiences are more common than you might imagine. In 2013, an enterprising South Carolina company launched a website called DiedInHouse.com. For twelve dollars, prospective home buyers can enter any address in the United States and find out if someone was murdered there.

Comfort with animistic thinking may diminish as people mature out of childhood, but it increases again later in life. In one study, only 5 percent of younger adults categorized as alive at least one item from a list of natural objects, such as mountains and oceans, compared to 30 percent of older adults.[4] Older adults tend to develop a distaste for mass-produced objects made of artificial material. They prefer things that are well used, one of a kind, and handmade from natural materials. These are the things of poetry, the kinds of things that take on supernatural properties most easily. An old wrought-iron fence feels more magical, more alive, compared to a brand-new plastic one.

Children adopt a magical outlook because they haven't fully developed their egos and remain under the influence of their unconscious. Young adults have more developed frontal lobes, and they've been schooled in discipline, self-control, and rational ways of thinking, all of which help them establish an independent ego. But to prevent that newly liberated ego from being overwhelmed by the unconscious, they need strong walls, which also may make them less susceptible to animistic experiences. The elderly, by contrast, have had a lifetime to build up ego strength, so it's safer for them to lower those walls and open channels of communication. This allows them to reexperience the magical world of childhood, but from a more sophisticated perspective.

In addition to things that are old or handmade, things that move can seem magical because they remind us of living creatures. In the movie *Speed Racer*, veteran race car driver Rex Racer coaches his younger brother as he drives around a track. Rex explains, "This ain't no dead piece of metal. A car's a living, breathing thing, and she's alive. Feel it talking to you, telling you what she wants, what she needs. All you gotta do is listen." French yachtswoman Isabelle Autissier once expressed something similar regarding her relationship with her vessel: "I spent three years . . . with this boat and I am very close to her. Now I am here, and she is alone and that is pretty difficult."[5]

Trees also deserve a place on the list of things that feel as if they're endowed with spirit. Trees often symbolize magic itself, and have long played a special role in folklore and art. Many fairy tales begin with the hero leaving the comforts of village life and entering the woods. It's a journey that takes him out of the ordinary world into an ancient, unknown world of enchantment. In the Brothers Grimm collection of fairy tales, we find enchanted pear trees, apple trees, hazel trees, and junipers. The legendary Tree of Life can be found in many sacred traditions, including the Jewish Kabbalah, Buddhism, Chinese and Norse mythologies, and African folklore.

On a more prosaic level, you may have seen stickers on paper towel dispensers that say "Remember . . . These come from trees." According to Peter Kazanjy, who designed the stickers, each one reduces paper towel use by more than eighty pounds every year. This is despite the fact that 90 percent of trees used to make paper are grown as crops, just like cotton and soybeans. No one gets upset about those plants being mowed down. But the idea of cutting down a tree, whether it grew in a forest or on a farm, gives us

Trees are different from other crops grown on farms. *(Source: Peter Kazanjy via Flickr.)*

a pang of momentary pain. We instinctively feel that trees are more than just a collection of molecules, and that they don't like being cut down. From a materialist standpoint, that's nonsense. But we should pause before discarding this irrational feeling. Our gut instinct that we have a special relationship with trees is supported by scientific evidence. An analysis of the mortality rates of more than eight million people compared those who lived in neighborhoods that were within half a kilometer of greenspace with trees to those whose neighborhoods were surrounded by buildings. The researchers found that the people who lived near trees lived significantly longer.[6]

Prior to the greenspace study, attributing special powers to trees would have seemed like childish superstition. But the irrational unconscious has a wisdom that goes beyond the understanding of the ego. It's important to test our instinctual feelings with rational deliberation, but we have to take a sophisticated approach. From an objective point of view, trees are just another plant. Cutting them down should be no different from harvesting soybeans. But here is a case in which accepting the magical perspective of the unconscious gives a person an actual survival advantage.

One might argue that the greenspace study eliminates the need to rely on magical thinking to explain the feeling that trees have a mystical connection to humans. That may be true, but it's worth noting that this study was only published in 2019. Science has a lot of catching up to do. Additionally, from the perspective of living life to the fullest, is it better to rely on columns of data to understand oneself, or is there an advantage to embracing the magical feeling that comes from the unconscious? We experience the world differently when we resist the mystical affinity we feel for trees in favor of relying on a sterile statistical analysis. We may avoid certain errors, but is it worth the price?

Fire is another example of something that easily becomes endowed with spirit. It moves in unpredictable ways and appears to pursue a goal—to burn up as much as possible. Firefighters have been known to describe massive forest fires as "devious," "cunning," or "lying in wait."[7] When the wind dies down, and the fire temporarily retreats, it's described as "resting up." Even ordinary objects can feel malicious. The lock that refuses to open carries a grudge against you, as does the vending machine that ate your quarters. Anytime you experience the urge to hit something that isn't behaving properly, it comes from the half-conscious belief that its behavior is intentional.

At the other end of the spectrum from malicious objects are objects that feel as if they have a soul because of our love for them. Typically, they're things we've owned for many years, such as a keepsake, a favorite quilt, or a trusty old paring knife. The longer an object is in contact with a human being, the more likely it is to be perceived as living. We value it not for its usefulness, but for its essential being. It's almost as if human minds radiate soul the way our bodies radiate heat. The many hours your mother spent crocheting the sweater she gave you, the love and care she put into it, make you feel as if she imbued it with life. If it were replaced by a machine-made sweater, even an exact replica, there would be a loss. Something magical would be missing—the most valuable part of the sweater.

AMBIENT SPIRITS

The spirits that bring everyday objects to life aren't the only ones we'll need to understand as we explore the unconscious mind. There are others as well, such as spirits that animate groups of people. Sometimes a crowd is more than a collection of individuals and takes on a life of its own. The feeling you get at a rock concert is different from the one you get listening to the same music at home. There is a divine frenzy that unites the crowd. The same thing happens in a mob that tears apart a city; its violence crosses the boundary of what any individual would do on their own as the moral restrictions of everyday life are overwhelmed by the spirit of the group. The will of the crowd is substituted for the will of the individual—and the switch is only noticed later, when calm is restored. Afterward, people may ask themselves, "What did I just do?" Even when only a few friends gather, the conversation can become so "spirited" that it feels like there's another person in the room, a mysterious presence driving things forward.

In addition to being present in things and groups of people, spirits can attend moments in time. The goddess Nike, the winged personification of victory, is the spirit of the magical moment when victory is assured. Athletes improve their skills through practice and training, but they can't control whether on any given day they will perform badly or well. They never know when everything will come together and they will deliver an inspired performance, one in which every shot hits its target, every move is performed with confidence. Even amateurs experience Nike's touch. When it happens, you feel as if you can't do anything wrong. Your victory is predestined. One of the most remarkable aspects of the experience is that it feels effortless. It's not something you're struggling to make happen. It's something that's happening to you, a gift delivered by an invisible agent.

Places, too, can sometimes feel as if they have a personality, as if there were a friendly spirit hovering in the air. The Romans called it *genius loci*, or "spirit of place," and it refers to a location's distinctive atmosphere. Today, we might call it a "vibe." Imagine the feeling that pervades the space around a cushioned bench in a bay window where you used to read as a child. Think of the hushed vibrancy of a leafy forest or the violence of ocean waves crashing on the sand during a storm. Antoine de Saint-Exupéry, the author of the modern fairy tale *The Little Prince*, invokes the *genius loci* of the nighttime desert when he writes, "One sits down on a desert sand dune, sees nothing, hears nothing. Yet through the silence something throbs, and gleams."

Animism may not be the central focus of our lives today, but in preindustrial societies an animistic understanding of the world was common, and being on good terms with the resident spirits was thought to be essential for survival. American literature professor Joseph Campbell, who is most famous for his work on the hero's journey, quotes an Apache storyteller who said, "The plants, rocks, fire, water, all are alive. They watch us and see our needs. They see when we have nothing to protect us and it is then that they reveal themselves and speak to us."[8]

It's not immediately obvious how the Apaches' needs are met when the natural elements reveal themselves. Perhaps the elements bring knowledge that helps the Apaches live in greater harmony with their environment. But there may be another kind of benefit. Campbell points out that the Buddhists call the phenomenon of natural revelation the "sermon of the inanimate."

There's a simple example of the sermon of the inanimate in the Disney film *Alice in Wonderland.* A patch of flowers sings a song called "All in the Golden Afternoon," in which we find the line, "You can learn a lot of things from the flowers." This sentiment seems innocuous enough, and if you've seen the movie and heard the song, you probably didn't give it a second thought. That alone is significant, because if you reflect on it for a moment, it's not at all obvious what they're talking about. Presumably, the flowers aren't referring to botanical knowledge, such as how many petals a daisy has. They seem to have something else in mind. But what? The answer feels both obvious and opaque at the same time. As we'll see, one of the difficulties with this question—what the sermon of the inanimate is trying to tell us—is that the answer can't be expressed in words.

Most modern people no longer live among plants, rocks, and trees. Technology insulates us from the close connection between our survival and our relationship with nature. Yet animism continues to work its magical influence over us. Look around your house. How many things do you see that you would refuse to trade for a newer, better model? How many things are valuable to you because of something invisible, something immaterial? Modern adults are contradictions when it comes to magic. We're conscious materialists and unconscious animists. We offer up denials when asked about belief in spirits. But throughout the course of daily life we contradict this professed disbelief in dozens of small ways, often without realizing we're doing it. The kinship we feel with an old sweatshirt exists on a different plane of awareness, like an old habit we forgot we had. But sometimes an encounter with a spirit is so powerful, we can't do the mental equivalent of sweeping it under the rug. The veil of the ordinary is ripped away, and the sermon of the inanimate is spoken.

THE NUMINOUS

The Kandinsky quote that opened this chapter reminds us that the souls in the things around us are silent more often than they speak. The friendly connection we feel with these objects exists as a background hum that enriches our lives but is brought to our attention only occasionally. There are times when an encounter with a spirit is more than just noticeable, however; sometimes it has a profound—even life-changing—effect. German theologian Rudolf Otto named this the "numinous experience." When it occurs, we feel that we are in the presence of something utterly different from the ordinary world. It's so different from what we're used to that Otto described it as *sui generis*, a Latin term meaning "in a class by itself." It's an uncanny feeling that some have compared to the shudders you get when you hear a ghost story, but rather than being frightening, it's a deeper, more uplifting experience.

Although the sui generis nature of the numinous makes it hard to describe, there are a few things we can say about it. First, we experience the numinous in a way that's different from anything else. Ordinarily, the only way we encounter something external to ourselves is via the senses. To perceive something, we must see it, hear it, smell it, taste it, or touch it. We see another person, hear her voice, or feel the grip of her handshake. But numinous experiences are a strange exception to this rule. Otto describes it as a "non-sensory experience or feeling whose primary and immediate object is outside the self."[9] Without sensory impressions to characterize it or words to describe it, the only way to understand it is to experience it.

Numinous experiences may not happen to everyone, but they're not rare. William James writes, "Our normal waking consciousness, rational consciousness as we call it, is but one special type of consciousness, whilst all about it, parted from it by the filmiest of screens, there lie potential forms of consciousness entirely different."[10] We experience these non-ordinary states of consciousness when the conscious mind is invaded by unconscious contents. It's most likely to happen when a person is alone, typically in a natural setting, such as the beach at sunrise. Or imagine being in the woods. The rays of the setting sun shine through the trees. The wind touches your face. Its fragrance reminds you of something you can't quite grasp, something you remember from long ago. Suddenly, it's no longer trees, sunlight, and wind. It's a single thing, invisible but alive, trying to tell you something.

A numinous experience is often *ineffable*. It can't be described in words; it must be experienced directly. In this respect, it's like falling in love. It's not possible to understand what it feels like unless one has experienced it firsthand. For those who never have, descriptions of it sound exaggerated or absurd.

It's also *familiar*. The feeling we're experiencing, the *entity* that makes its presence felt, is something we used to know, but lost touch with. As author G. K. Chesterton writes, "All that we call spirit and art and ecstasy only means that for one awful instant we remember that we forget."[11]

The numinous is *personal*. It knows who you are and wants to communicate something that no one else would understand. It's *invigorating*. It makes you feel wide awake, more alert than you are under ordinary circumstances. It's full of vitality, simultaneously urgent and serene. And it stimulates *yearning*. When the feeling is gone, you experience an intense longing for it to return. The ordinary world is bland in comparison.

William Butler Yeats writes about yearning for the numinous in his poem "The Song of Wandering Aengus." The narrator describes a magical experience that occurred while fishing in the woods at night. He catches a silver trout that transforms into a shining female figure.

It had become a glimmering girl
With apple blossom in her hair
Who called me by my name and ran
And faded through the brightening air.

The vision moves him to spend the rest of his life trying to find her again, which he believes will gain him entry into a magical paradise.

Though I am old with wandering
Through hollow lands and hilly lands,
I will find out where she has gone,
And kiss her lips and take her hands;
And walk among long dappled grass,
And pluck till time and times are done,

The silver apples of the moon,
The golden apples of the sun.

The theme of the numinous lover is timeless. A contemporary description can be found in the popular song "Sweet Child o' Mine" by American rock band Guns N' Roses. The first verse describes the smile of the artist's lover as a gateway back to an idyllic childhood, "when everything was as fresh as the bright blue sky."

Scientists have also taken an interest in the numinous. Frederick Barrett and colleagues developed a scale to measure its intensity called the Mystical Experience Questionnaire, which contains thirty statements describing feelings that may occur during an encounter with the numinous. Respondents rate each statement on a scale of 0 to 5 to indicate whether they had the experience and how strong it was. When researchers asked 184 volunteers who had been given psilocybin, one of the ingredients in "magic mushrooms," to fill out the questionnaire, they found that higher scores on the questionnaire predicted more enduring changes in the respondents, including changes in attitude, behavior, and well-being. The stronger the mystical experience, the more profound were its lasting effects.[12]

Sample Items from the Mystical Experience Questionnaire

Feeling that you experienced eternity or infinity.

Feelings of peace and tranquility.

Experience of the insight that "all is One."

Feeling that you experienced something profoundly sacred and holy.

Awareness of the life or living presence in all things.

Feeling that it would be difficult to communicate your own experience to others who have not had similar experiences.

Items on the Mystical Experience Questionnaire are rated from 0 (*I did not feel this at* all) to 5 (*I felt this more than any other time in my life*). Mystical experiences can occur during religious rituals, drug intoxication, and in the course of ordinary life. These experiences often lead to enduring changes in the way people see themselves and the world.

One of the questionnaire items refers to the mystical revelation of universality, the realization that all is One. It's a mystical feeling that many of the study volunteers experienced but that doesn't require psychedelics to evoke. A different survey of people who reported mystical experiences that were spiritual in nature also found high levels of universality. These feelings of universality were related to how real the experience felt compared to usual reality.[13] It's not easy to imagine how something could be more real than reality, but 70 percent of the respondents said the experience was either "somewhat more real" or "much more real." It's one of those ineffable qualities of the experience, and is frequently accompanied by the feeling of being most fully alive when we connect with our unconscious. When the researchers asked the respondents to write descriptions of the experience, words that reflected universality were more common in the descriptions written by those who perceived their experience as "more real than usual reality."

The researchers found that high levels of perceived reality were also associated with positive impacts on family life, health, and sense of purpose, and a reduced fear of death. The numinous experience may be alien, representing a dramatically different way of perceiving the world compared to ordinary states of consciousness, but it's not pathological. It brings important benefits that can enrich our lives.

Less Real	More Real
I	we
might	must
not	all
or	and
probably	everything
think	love

Words statistically related to high and low levels of perceived reality during a mystical experience. Words related to universality were more common when the experience was perceived as "more real than usual reality."

THE SCIENCE OF ANIMISM

We tend to be more effective when we assume that things around us have conscious intentions and goals.[14] That may sound strange, because we associate this kind of belief with children and preindustrial cultures. But attributing consciousness to nonhuman objects increases our ability to make sense of their behaviors. Even scientists indulge in this kind of thinking because it helps them organize and interpret their data. Charles Darwin was known for describing the behavior of animals in terms of human emotions. He described dogs as showing affection, making friends, and dearly loving their offspring.[15] This type of description is called *anthropomorphism*, which means "in the shape of humans."

Some scientists have criticized anthropomorphism as a way of understanding animal behavior. Peter Broadhurst, a psychologist who worked at the Stanford University Center for Advanced Study in the Behavioral Sciences, called it "the cardinal crime for the animal observer."[16] He writes, "There is little justification and less explanatory value in ascribing man's elaborate conscious processes to animals and discussing emotions in such terms would be futile." Based on this apparently reasonable criticism, researchers who studied chimpanzees at the Yerkes National Primate Research Center decided they would avoid all anthropomorphic descriptions for two years to see if that would improve the quality of their observations. They were surprised by what happened:

> All that resulted was an almost endless series of specific acts in which no order or meaning could be found. On the other hand, by the use of frankly anthropomorphic concepts of emotion and attitude one could quickly and easily describe the peculiarities of individual animals . . . Whatever the anthropomorphic terminology may seem to imply about conscious states in chimpanzees, it provides an intelligible and practical guide to behavior.[17]

When they avoided anthropomorphism, they were able to collect data, but they couldn't turn that data into knowledge because they couldn't conceptualize it in a meaningful way. The notebooks full of observations didn't help them understand the behavior of the chimpanzees until they allowed themselves to think about the chimpanzees in human terms, so that their behaviors became stories.

We have two ways of understanding: an instinctual one, which views the world from a magical perspective, and a rational one, which tends to be more materialistic. The instinctual way of understanding is the one that comes naturally. When we watch the behavior of animals, we unconsciously assign human-like motivations to their actions. We don't have to use our imagination. We don't have to ponder or deliberate. It's a built-in feature of the brain that happens outside conscious awareness.

The rational approach is different. It's a conscious effort to test the accuracy of our instinctual beliefs and decide if they're valid. In this case, we know that animals are different from humans, and that an anthropomorphic understanding will lead to errors. The logical thing to do, therefore, is to reject our instinctual anthropomorphic approach. Yet the rational approach hampered the scientists at Yerkes so much, they were unable to constructively interpret their observations.

Where does that leave us? Ideally, we'd like the two ways of thinking to work together, but that's hard. The logical part of the mind is something of a dictator. It doesn't like working with the instinctual side because, in its estimation, the instinctual side is simply wrong. The instinctual side can also be uncooperative. It operates outside of conscious control, doing what it wants, when it wants, and unlike logic, it won't behave as an obedient servant. Most people prefer the logical approach, but as the Yerkes researchers found, you pay dearly if you ignore a large part of your brain and try to make sense of the world using only half your cognitive abilities.

It's not only zoologists who find it useful to allow intuition to mingle with logic and portray their objects of study in human terms. Physicists who have difficulty finding an elusive particle call it "shy." Chemists describe molecules as having "affinities" for one another. Doctors treat "stubborn" infections and "aggressive" tumors. Using anthropomorphism to understand natural phenomena is pervasive in science, even at its highest levels.

Why is it so much easier to manage information in a social context (one conscious being interacting with another) as opposed to a mechanical one (a conscious being interacting with a nonconscious object)? Processing complex social information is one of the most important functions of the brain. It involves some of the most sophisticated neural circuits humans have, and large areas are devoted to this task. Processing this information is so central to the way we think that some scientists believe the evolutionary benefits of being able to manage complex social structures are the reason *Homo sapiens* developed such large brains.[18]

Cooperating with others gives us access to food, shelter, and tools that we wouldn't be able to obtain on our own. As we've seen, cooperation provides an important evolutionary advantage, which is perhaps why the need for social interaction is universal, seen across all cultures, and present from the moment of birth. It's such a basic human need that when we don't experience frequent contact with other people, we can develop serious health problems—even die. Among older adults, social isolation is as likely to lead to an early death as smoking fifteen cigarettes a day.[19]

One of the most important skills required for human cooperation is the ability to understand what someone else is thinking—to be able to put yourself inside the head of another person. This ability is unique to human beings, and we're so good at it, it's effortless. You probably have no difficulty understanding the following sentence: *I could tell that my sister and I were thinking the same thing: what our parents really wanted was for us to be happy.* But consider: to grasp the meaning of that sentence, you must put yourself in the mind of the narrator, who is putting herself in the mind of her sister, who is putting herself in the minds of her parents, who are putting themselves in the minds of their children. That's an enormously complicated task. Experiments with chimpanzees indicate they can't take even the first step. But the human brain is so well suited to this type of processing that we don't even realize how hard it is. In fact, we find it easier than the computationally much simpler task of multiplying 15 × 15. It's like a computer that has no difficulty displaying animated 3D graphics because it has a sophisticated graphics processing unit. Human brains have sophisticated social processing units.

When we try to decode the thoughts and intentions of another person, the most important source of visual information is the face. Newborn infants orient their gaze to their mother's face, and faces remain the most conspicuous part of our environment throughout life. There's a part of the brain, located in the temporal lobe, that's so specialized for understanding faces, it's

The fusiform face area of the brain can be activated by nonhuman objects if we have a strong connection to them. ***(Source: Cleanpng.)***

been named the "fusiform face area."[20] It's what helps us extract social information from facial expressions, and its circuits shift into high gear whenever faces appear in our visual field.

Yet, under certain circumstances, this highly specialized area also responds to non-face objects. Viewing objects in one's area of expertise also activates this brain area.[21] When an auto mechanic looks at a car, his fusiform face area springs into action as if he were seeing a human face. The same is true when an ornithologist views a bird or a lepidopterist sees a butterfly. As one develops an intimate familiarity with nonhuman objects, the brain begins to treat them as human.

An understanding of the fusiform face area helps us solve an apparent paradox: the more time you spend with an object, the more likely you are to experience it as more than just atoms and molecules. It's counterintuitive because you would think that the better you know an object, the more accurate your understanding of it would become. If a paring knife is nothing but metal and wood, a longstanding familiarity with it should lead to an increasingly materialistic judgment regarding its true nature. But, in reality, the opposite occurs. The longer a chef uses a particular knife, the more likely she is to experience it as having a soul. Over time, her brain shifts away from circuits that process inanimate objects and begins to favor circuits that process social relationships, such as the fusiform face area. The paring knife is no longer an ordinary tool; it has become a friend.

THE BENEFITS OF SEEING SPIRITS EVERYWHERE

Acknowledging our biological tendency to see a world full of spirits is not something we're used to doing, but it has advantages. First and foremost, it's not a good idea to fight your own biology. You wouldn't want to spend the rest of the day walking on your hands instead of your feet. Similarly, willfully opposing brain circuits that promote a magical outlook may not be the most effective path to a fulfilling life. Imagine an amateur carpenter shaping a piece of wood with a hand plane that belonged to his father. He might choose to deny the feeling that the plane gives him a special connection to someone who played an important role in his life. "It reminds me of him," he might say. "Apart from that, it's no different from one I could pick up in a store." But then he would miss out on the experience of emotionally connecting to the memory of his father in a way that wouldn't be possible without the magical belief that the plane contains an echo of his father's spirit. Or think of the chef who has a social connection with the paring knife she bought while in culinary school. How might having a friendship with this essential tool affect the way she prepares food—and therefore the quality of the dishes she makes?

A survey of 200 people with an interest in the supernatural found that supernatural beliefs had positive effects on their lives.[22] Three-quarters felt more comfortable and peaceful when they thought about their beliefs; 86 percent said these beliefs helped them understand themselves better, particularly with regard to their emotions. Like magical beliefs, emotions are the result of unconscious processing. Acceptance of one (magic) appears to lead to a better understanding of the other (emotions). When you open the passages to the unconscious parts of your brain, lots of things come through that door.

On the other hand, when we discount this way of thinking, we limit ourselves in numerous ways. For one, we don't perform as well in situations in which there is both uncertainty and high stakes. Just as the participants in Subbotsky's studies became more open to magical ways of seeing the world when he raised the stakes, people often turn to sources of supernatural help when faced with an exam, job interview, or first date.[23] The friendly spirits provide comfort and boost confidence, which in turn can enhance our abilities.

For professional athletes, more than almost any other group, achieving peak performance is essential to a successful career, and they instinctively know that a magical approach gives them an edge. It comes in the form of superstition. Athletes must constantly cope with uncertainty; they never know when the goddess Nike will grace them with her touch, or when the ball will take an unexpected bounce, sending them scrambling. Superstitious rituals allow them to regain a feeling of control. Athletes are generally more superstitious than other people, and professional athletes more so than amateur ones. Elite athletes are the most superstitious of all.

Michael Jordan had a pair of practice shorts from his college basketball team, the University of North Carolina Tar Heels, that he always wore underneath his Chicago Bulls shorts. Tiger Woods always wore a red shirt during tournaments. NFL quarterback Tom Brady carried "protection stones" when he played, which were given to him by his wife, Gisele Bündchen.[24] These lucky charms boost confidence, and in sports, increased confidence translates into better performance.

When research volunteers were asked to make ten putts from a distance of 100 centimeters, those who used a "lucky" ball performed better. When these volunteers were handed the ball, they were told, "Here is your ball. So far it has turned out to be a lucky ball." Other volunteers were given the same ball but were simply told, "This is the ball everyone has used so far." The putters who got the "lucky" ball sank an average of 6.42 shots, compared to only 4.75 for those who used the ordinary one.[25]

NCAA bowling champion John Hinkle Jr. tapped into the magic of the lucky ball in a particularly dramatic way. After his father died, Hinkle filled a ball with his father's ashes and was able to bowl a perfect game. He said the experience gave him goosebumps, and by the last frame he had so many tears in his eyes he couldn't see where the ball was going.[26]

Our interactions with the spirits that inhabit the things we love are the numinosity of daily life. Yeats's wandering Aengus spent a lifetime pursuing his numinous vision, but perhaps it was inside him the whole time. Natural selection has arranged the circuits of the human brain to discern glimmers of the numinous all around us, in otherwise ordinary objects. These objects may be silent more often than they speak, but if we listen, we will hear them.

Life is difficult, and we need help. Our helpers are the spirits that have embedded themselves in the wiring of our brains. When we push these helpers away, and deny their influence over us, we are impoverished. Yet, we do it anyway, because they don't fit the quasi-materialism of the modern world. So let's take a closer look at what modern science says about the strange division that exists within the human mind. By looking more closely at the scientific understanding of the unconscious, we'll gain a better understanding of how supernatural spirits shape our lives.

THE UNCONSCIOUS IN THE LABORATORY

The intuitive mind is a sacred gift and the rational mind is a faithful servant. We have created a society that honors the servant and has forgotten the gift.

ALBERT EINSTEIN

THE ROAD TO FAIRYLAND

One of the first things Sigmund Freud noticed as he began to treat his patients was that, in many cases, the symptoms they complained about appeared to be volitional—choices patients were making of their own free will. The symptoms weren't like the paranoid delusions of schizophrenia or the irrational despair of clinical depression. Delusions and despair are outside a person's control, like chest pain or difficulty breathing, which makes them easier to understand.

What puzzled Freud were patients who complained of not being able to stop washing their hands or bingeing on food. Why would you see a psychiatrist for this type of problem? If you don't want to wash your hands so much, stop washing your hands. If you don't want to binge, stop eating. But, of course, his patients couldn't. That's why they needed a doctor's help.

Freud concluded there must be forces within the brain that are stronger than volitional will. This realization, that forces within the brain conflict with one another, led him to his theories of the unconscious. Today, his approach is called "psychodynamic," because it focuses on the dynamic tension among these psychological forces.

By closely observing his patients, looking for clues that would help him understand the battles going on inside them, Freud learned a great deal about the unconscious and how internal conflict can lead to symptoms. But while observation is the essence of science, observing patients under one's care isn't the most scientifically rigorous form. Observation produces anecdotal evidence, which can get a scientific discipline started, but better types of evidence can be collected by observing research volunteers more systematically.

Scientific research is designed to be objective, adopting an investigative approach that looks at the human mind from the outside. A subjective approach, by contrast, takes advantage of the unique perspective we have from the inside of our minds—a perspective that allows humans to share insights into our inner world through literature and other forms of storytelling.

These subjective descriptions may be the richest source of information we have about the unconscious. Moreover, scientists attempting an objective approach to studying the unconscious must overcome some formidable obstacles. If we want to know a person's conscious thoughts, we can ask him. We can do the same with opinions, attitudes, or anything else that's conscious. But how do we study something a person is unaware of? By definition, the unconscious can't be perceived. It has a mind of its own, and it acts in its own way, in its own time.

Another challenge scientists face is that human behavior arises from the combination of conscious deliberation and unconscious influences. How can you eliminate the role of the ego to make sure you're measuring only the unconscious?

With these caveats, it's still best to begin by looking at what objective studies tell us about the unconscious and how it gives rise to magic. The road to fairyland passes through the laboratory.

WHISPERING SOFTLY IN THE EAR OF UNCONSCIOUS

Scientists have used several approaches to address the challenges of peering into the hidden places of the human mind. One is to communicate ideas in a way the conscious mind won't perceive but the unconscious will. It's like whispering in someone's ear so quietly, only they will hear. If we can do that, we can take the ego out of the equation and be sure that any reaction comes straight from the unconscious and slips past the ego. As it turns out, the brain is organized in a way that allows us to do just that.

To understand how this is done, start by looking around. If you're in a well-lit room, you can see everything in your surroundings. The same is true of your other senses. You can smell the odors in the air, hear the sounds, and feel the textures of the things you touch. The world is an open book, flowing into your conscious mind through your senses.

Except your brain doesn't actually work that way. Your conscious mind gets only a small part of the information that enters through your sense organs, and that information is so highly processed that the picture of reality it presents is not always accurate. Your ego is like the offspring of wildly overprotective parents who have shut their poor child inside a box. All the child knows about the outside world comes from what his parents slide in through a small hole. Like all overprotective parents, they generally have his best interests at heart, so they give him information they think will be useful to him. But before he gets it, they censor it, filter it, and process it into a form they believe his young mind will be able to handle.

That's the way the unconscious treats the ego, and although it sounds offensive, we should be grateful. Remember, the ego can only process about 60 bits per second. The amount of data flowing through the senses would overwhelm the ego unless the data were filtered and processed. The unconscious makes decisions about what it thinks the ego should be paying attention to and suppresses everything else. Think about what it does to your sense of touch. If you run your fingers across the page of a book, you experience the texture of the paper. But while you're doing it, you're completely unaware of the sensation of your clothes touching your body and the pressure of the floor on your feet. If you were constantly bombarded by those sensations, you wouldn't have any bandwidth available to pay attention to other things that are more important.

An example of what happens when the unconscious doesn't do a good job filtering sensory input is attention-deficit/hyperactivity disorder (ADHD), a mental illness characterized by poor focus and concentration. People with ADHD have trouble focusing because they don't block out irrelevant stimuli. They may be sitting in their office, reading a report, and lose their train of thought every time someone walks by their door. Overhearing a conversation or simply a car driving down the street outside their window can be enough to distract them from what they want to concentrate on. The opposite of this is a person who concentrates so well that if she's immersed in a book, she may fail to notice someone in the same room calling her name.

In general, it's a good thing to have the unconscious create a sort of barrier around us—a wall of isolation, so to speak—to prevent irrelevant stimuli from getting through, so the ego can focus on what's important. If you're talking to someone in a crowded room, you can hear and pay attention to what your conversational partner is saying because all the other conversations are suppressed. If you're walking down the street at night, and a black cat darts across your path, it doesn't matter if it's barely visible in the darkness. Thanks to unconscious filtering and processing, the shadowy figure becomes the most prominent object in your visual field. The light bouncing off the license plate of a car parked under a streetlamp would also enter your eye, but only your unconscious would perceive it. It would be hidden from the ego.

What if we embedded a message within the letters and numbers of the license plate? Could we communicate with the unconscious behind the ego's back? Could we use that communication channel to learn more about how the unconscious behaves and what it's capable of?

A strategy much like this was used in a series of experiments designed to explore unconscious agency.[1] Earlier, we looked at ways in which unconscious agents develop independent goals that are hidden from the ego. One example we looked at was the "thirsty demon" in the brains of alcoholics. It can be uncomfortable to imagine there are hidden agents within our brains who are pursuing goals outside our awareness. It's so alien to the way we think about our own minds, it's hard to accept in the absence of strong evidence. So, a group of researchers designed a series of experiments to look for that evidence—to see if they could smoke out these agents and catch them in the act of pursuing goals outside the awareness of research volunteers.

Imagine a group of people attending a lecture on the health benefits of eating nuts. We wouldn't be surprised if we saw an increase in nut consumption in the days and weeks following the lecture. This is an example of the stimulation of conscious goals. Conscious goals have a number of predictable characteristics. The first is very simple: conscious goals are pursued. These goals lead to behaviors that make the realization of the goals likely—in this case, driving your car to the grocery store to buy a bag of walnuts. Second, people demonstrate persistence when they have a conscious goal. If they encounter obstacles, they don't give up. For example, if the road leading to the grocery store is blocked off for repairs, you take a detour. Last, conscious goals are pursued even if they're interrupted. If you planned to buy walnuts this afternoon, but you have to work late, you go tomorrow or over the weekend instead. A goal doesn't fade away like a passing fancy. We pursue it until it's accomplished. The question the researchers wanted to answer was, is the unconscious the same? Does it pursue goals, overcome obstacles, and maintain its intention despite interruptions?

Their first experiment was designed to test whether unconscious agents will pursue goals on their own initiative. Volunteers played a fishing game that allowed them to either compete against their opponent to gain the highest possible profit or cooperate for the good of all by giving up profits to restock the lake.

The volunteers were divided into three groups. The scientists gave the first group no instructions with regard to cooperating or competing. They gave the second group explicit instructions to cooperate by throwing fish back into the lake to restock the supply. They instructed the third group to cooperate, too, but in a way that only the unconscious would perceive.

Here's how the researchers slipped the message past the ego. Before playing the fishing game, all the volunteers were tricked into believing they were performing a "psycholinguistic task" in which they had to make sentences out of sets of words. The first two groups were supplied with neutral words, like *salad* and *umbrella*, while the third group was given words such as *dependable*, *honest*, and *fair*. When all aspects of the experiment were complete, the researchers went back and interviewed the volunteers to see if they had noticed anything suspicious about the psycholinguistic task. None of the participants gave any indication of being aware of what had taken place. The "whisper" worked. Their conscious minds didn't notice what was happening.

When it was time to play the fishing game, the players who received no instructions returned an average of twenty-five fish to the lake. Those who were explicitly told to cooperate returned almost 30 percent more. And those who were given the secret instructions, which only their unconscious heard? They threw back the same number of fish as the explicit group. They intuitively felt cooperation was the right thing to do.

Similar experiments found that unconscious goals continued to be pursued when researchers created both delays and obstacles that had to be overcome. Secret messages passed to the unconscious resulted in volunteers persevering despite (intentionally) broken equipment, opportunities to switch from difficult tasks to easy ones, and even explicit instructions to stop working. The volunteers had no idea why they were so persistent; they just kept going.

THE POWER OF NOT THINKING

Once we establish the existence of unconscious thought, the next step is to better understand its capabilities. Does it function the same way as conscious thought, or does it approach problems in a different way with different strengths and limitations? If there are significant differences, it's important to find out what they are. To get the most out of our brains, we need to know what the ego should take responsibility for, and what it should pass along to the unconscious. Also, if we want to hand off a task to the unconscious, how is it done?

To answer these questions, we'll look at a new strategy researchers developed to separate the thought processes of the ego from those of the unconscious. This strategy exploits the limited processing power available to the ego. If we present a research volunteer with a problem to solve, then distract her ego with a task that uses up all its processing power, only the unconscious will be able to work on the real problem. The solution that emerges will be the result of unconscious thought, and we can then compare this solution with one found by an undistracted volunteer who's able to use his ego to work on the problem.

The researchers who designed these experiments began by making predictions about what kinds of problems the ego would be best at. The ego excels at reasoning and following precise instructions, but since it has only limited processing power, it should do best with straightforward problems that don't involve too many variables. The unconscious, with its massive processing power, ought to be better at solving more complex problems. However, because it doesn't use logic, one would expect its answers to be more impressionistic. If you were thinking about investing in a new business, it might make sense to ask your unconscious whether your potential business partner seems honest, but you'd want your ego to calculate the internal rate of return you can expect on your money.

Consider the following problem. You're moving to a new city, and you're looking for a house. You'd like one close to your work and big enough for a growing family, and it can't cost more than $350,000. Your agent shows you three homes. The first is a beautiful, large house ten minutes from your office that costs $425,000. The second is almost large enough, is located thirty minutes from your office, and costs $300,000. The third has plenty of room, is located fifteen minutes from your office, and costs $335,000.

Your conscious mind would have no trouble making that choice. There are only a few variables and the rules are clear. But what if things were more complicated? Researchers at the University of Amsterdam asked volunteers to consider four apartments and choose which one they thought was best.[2] Each apartment was described by twelve different features, which meant the volunteers had to sift through a total of forty-eight pieces of information that were presented in random order. The features were set up so that one apartment was objectively better than the others, one of them was worse, and the other two were in between.

The participants were divided into three groups. One group had to make an immediate decision about which apartment was best as soon as they finished reading the descriptions—a snap judgment. Another group was given three minutes to think about it, and the final group was distracted for three minutes before they made their decisions. Participants in this last group were asked to solve anagrams, in which they had to rearrange letters to make words; for example, they had to make a word from the letters HACTW. Challenging anagrams use all the ego's processing power, so that, in this group, only the unconscious mind was able to think about the apartments.

The results confirmed that the apartment problem was a difficult one. The immediate choosers and the conscious deliberators were just as likely to choose the undesirable apartment as they were the desirable one. They were overloaded with so much information, it was as if they chose at random. The group that did the anagrams for three minutes, however, tended to choose the desirable apartment. Using distraction to take conscious deliberation out of the equation while the unconscious worked through the problem allowed it to find its way through the thicket of details and reach the correct solution. The volunteers' experience is reminiscent of the Schopenhauer quotation from the first chapter, on how, after familiarizing himself with the factual data of a problem and then not thinking about it for a few days, the answer came to his mind "entirely from its own accord."

A similar study measured consumers' satisfaction with their purchases of a number of common products.[3] Some products were simple, such as a bottle of shampoo, and some were complex, like a camera. Researchers in Amsterdam measured how much conscious thought went into the decision of which brand to buy. They found that with simple purchases, extensive conscious analysis of the various options led to greater satisfaction, while for complicated purchases, the opposite was true: more time spent thinking about which brand to buy led to lower levels of satisfaction.

This finding contradicts common sense. We tend to believe that complicated decisions should be preceded by a more thorough consideration of all the various features. But the ego isn't the only part of the brain able to form judgments. Sometimes the unconscious is better suited for the job. As the researchers note, "As decision makers, people are bad managers of their own minds. They behave like a conference organizer who asks the janitor to deliver the keynote address and the highly accomplished professor to fold up the chairs."

What about when there are no explicit criteria for what is desirable? The ego needs rules to make decisions. An absence of clear rules would be an indication that we're in the realm where the unconscious has an advantage over the ego, and a study that measured satisfaction with pieces of art confirmed this conclusion.[4] Participants were told they could take home one of five posters. Half the group were told to make the choice based on their gut feeling, and the other half were told to carefully scrutinize the reasons for their preference. They were given a few weeks to live with the posters, and then asked how they felt about their choice. Those who scrutinized the posters carefully were less happy than those who had made a gut decision. When there are no clear criteria for making a rational choice—as in this case, where it was simply a matter of taste—deliberation and introspection are counterproductive. Are you trying to decide what color to paint the kitchen or which shirt to buy? If you want to make the best decision, stop thinking about it, wait a little while, and then go with your gut.

As we've noted, the processing power of the unconscious is half a million times greater than what the ego has at its disposal. But the unconscious has another advantage as well. The logical ego solves problems based on what makes sense. It follows a linear train of thought, traveling the well-lit paths of rational deliberation. But the unconscious ventures into the dark corners of the mind where the ego doesn't go. The unconscious embraces irrationality, and this means it has access to a much larger database of possible solutions. Most of these irrational solutions will be useless, but that's all right. The unconscious has the bandwidth to scan through this extensive collection of options for the most viable ones, in the process often arriving at a solution the conscious mind never would have considered. That's the source of creativity.

A choice made by the unconscious comes to us in the form of intuition. Some people distrust intuition, especially when it comes to complex decisions. They equate it with a snap judgment. With both snap judgments and intuition, we choose without conscious deliberation, but, as we saw in the Amsterdam apartment study, they're not the same thing. A snap judgment is choosing with no thought at all, whereas intuition reflects thought that's outside conscious awareness. It can be hard to trust these hidden agents, but they have capabilities the ego lacks, and it's beneficial to know how to work productively with them. It goes without saying that intuition isn't infallible any more than rational deliberation is. It would be a mistake to expect the unconscious to always be right. But when it comes to weighing a large number of factors based on fuzzy criteria, the unconscious will find the correct answer more often than the ego.

There are times when we instinctively know it's best to hand off an important decision to the unconscious. When faced with a complex decision that will have long-term consequences, we often say, "Let me sleep on it." Sleep is when the unconscious assumes the dominant role in a person's mental life, and we know there's a good chance we'll wake up in the morning with a gut feeling that will point us in the right direction.

The ego and the unconscious are good at different types of mental processing, and humans function most effectively when there is an optimal division of labor. Working with the unconscious isn't always easy. It can be unreliable; we never know when it will speak and when it will be silent. But if we develop the habit of paying attention to our intuitions and gut feelings, we can cultivate a valuable partnership.

MAKING A MONSTER

As we've seen, it makes sense for the ego to step away from certain decisions and let the unconscious take over because it's better suited for the job. That's a sensible thing to do. But it's not such a good idea to dump things into the unconscious because they're unpleasant and the ego doesn't want to deal with them. If a person who considers himself compassionate realizes one day that he has been cruel to someone who is weak, that's a bitter pill to swallow. It might require him to reevaluate who he is and give up cherished notions of himself. Sometimes it's easier to deny reality. He might justify his actions by weaving together some pseudorational excuse, or he might conveniently forget what he did, banish the thought from consciousness, and push it down into the unconscious. What are the risks we face when we refuse to accept ownership of our feelings and actions?

Scientists at Trinity University addressed this question by introducing a potentially disturbing idea to a group of volunteers, then asking some of them to push the thought away.[5] Like the research done at the University of Windsor described earlier, the topic they chose was sex. The theme of sex comes up frequently in psychological research for two reasons. The first is that promoting reproduction is one of the main goals of evolution, making sex an effective subject for studying behaviors driven by instinct.

The second reason is that sex often leads to conflict between the ego and the unconscious. It's not uncommon for agents within the unconscious to be interested in sex with a wide variety of partners, both appropriate and inappropriate. The ego must step in, evaluate the urges the unconscious throws at it, and decide which ones to act on. It's not an easy business, trying to block the unconscious from pursuing its reproductive interests, and anyone who reads the news from Hollywood and the nation's capital knows that the ego's protests often fail to carry the day. But win or lose, sex creates conflict.

This conflict is more than psychological. It also triggers a physical stress reaction. We may not be aware of the stress, but our bodies still react to it in measurable ways. In this study, the research volunteers' stress levels were determined by how much they perspired. While we don't drip with sweat except under severe stress, even a tiny amount of salt water on the skin makes electricity flow more easily, which can be measured via attached electrodes. Something similar is done when a person undergoes a lie detector test. The technical term for this physiological response is skin conductance level or SCL.

First, researchers established that thinking about sex was more stressful than thinking about something else moderately stimulating, like dancing. As expected, thinking about sex raised SCL more than thinking about dancing. They also found that the effects of suppressing these thoughts showed the same pattern. When research volunteers were told not to think about sex, they perspired more than volunteers who were told not to think about dancing.

The Trinity University team's next experiment was complex, but its findings go to the heart of why suppressing disturbing thoughts may not be a good strategy for attaining peace of mind. The researchers told everyone to think about sex, then told half of the subjects to stop thinking about it, while the other half received no further instructions. Those who received no instructions continued to think about sex, but after a while, the topic faded from their minds. As it faded, their SCL settled back to normal. By contrast, those who were told to suppress thoughts of sex reported some initial success, but the idea kept returning, intruding into their thoughts uncontrollably. Each time it did, their SCL spiked upward.

When we permit ourselves to think about disturbing ideas, the conscious mind can process them and work through the associated difficulties. But if we disavow them, they lie in wait, raw and unprocessed. It's a serious matter. The monster we create by pushing disturbing thoughts into the unconscious does more than just stress us. It can kill us. People who have what's called a repressive coping style—those who manage strong emotions by dismissing or ignoring them—are more vulnerable to developing cancer.[6] The high concentration of stress hormones circulating through the body suppresses the immune system, which makes it harder to eliminate cancerous cells. People with a repressive coping style are also at greater risk for high blood pressure and other chronic illnesses.

Various therapeutic techniques attempt to clear away these toxic agents through the opposite of repression. One of them, called free association, is a technique Freud used when psychoanalyzing patients. Instead of repressing disturbing thoughts, patients made a commitment to report whatever came to mind, no matter how ridiculous, embarrassing, or upsetting. In this context, the word *free* means that the ideas are free from conscious censorship. Freud told his patients, "Act as though, for instance, you were a traveler sitting next to a window of a railway carriage and describing to someone inside the carriage the changing views which you see outside."[7] The assumption is that the most active and troublesome thoughts, when freed from the restraining influence of repression, will rise to the surface where they can be consciously processed, so they no longer have to be pushed into the darkness.

Some thoughts cause more than just discomfort. Memories of trauma, for example, can be overwhelming. The ego simply may be unable to process them because they are too strong. When that happens, it can lead to post-traumatic stress disorder (PTSD), a psychiatric illness that can occur after sexual violence, serious injury, or the risk of death and causes debilitating symptoms that are both physical and psychological. For people with PTSD, feelings of fear don't fade with time as they would under ordinary circumstances; instead, they experience an always-present sense of threat. At any moment they can be overcome by terror, even in safe environments. In some ways, it's like the recurring sexual thoughts the repressors in the SCL experiment couldn't get rid of, except the magnitude of trauma is so great that it damages the ego's ability to function.

A common form of psychotherapy for PTSD runs the process of repression in reverse. Called graded exposure therapy, it's an approach that helps patients address the trauma a little bit at a time by giving the ego only as much as it's able to process. If someone was assaulted in a dark alley, the first step might be for them to imagine walking down a street in bright daylight. Over time they progress to imagining a darker street, then the alley itself. Thanks to the ego's ability to process disturbing material and make it safe, the person gradually builds up the strength to replay the entire event in their imagination, allowing the ego to experience the memories without becoming overwhelmed.

As we delve deeper into the symbolism of the human psyche, we'll often find light used as a symbol of consciousness and reason. Light reveals things that are hidden, allowing us to distinguish one thing from another—the essence of analysis. Light also disinfects. If we take frightening thoughts and bury them in our unconscious, they fester and haunt us, popping up periodically to disturb our peace of mind and increase our stress hormones. Facing them head on is painful, but lifting them into the light of consciousness stops them from acting as autonomous agents.

TOP-DOWN VERSUS BOTTOM-UP

Consciousness is associated with the most recently evolved part of the brain, the prefrontal cortex. If you start at the bottom of the spinal cord and make your way upward to the point where it meets the brain, then continue through the subcortical regions, all the way to the area that lies just behind the forehead, the prefrontal cortex is the last stop on your journey. It's at the top. Consequently, when the conscious mind generates signals that affect the body or other parts of the brain, it's called "top-down" processing. Consciously moving your hand is one example. You also engage in top-down processing when you suppress reactions to emotions, staying calm in a crisis or mastering your anger. Bottom-up processing is the opposite. It represents signals flowing into the prefrontal cortex that originate in the body or in lower parts of the brain. Emotions and sensory information making their way up into consciousness are examples of bottom-up processing.

Nobel Prize–winning economist Daniel Kahneman describes this dual processing model of thought in his book *Thinking, Fast and Slow*. He calls bottom-up processing System 1 (because it's evolutionarily older) and top-down processing System 2. Rational, top-down processing is slow, requires effort, and is free of emotion. Bottom-up processing, which is also called "experiential processing," is fast, effortless, and doesn't follow the rules of logic. It's called experiential because this system is modified (i.e., learns) via experience rather than language (such as reading a book or listening to a lecture). In addition to producing sensory experiences and emotion, experiential processing also generates ideas that come to us in the form of intuition.

Rational, top-down processing understands the world categorically, using abstract concepts. It classifies things by taking them apart and focusing on specific properties, rather than looking at them as a whole. Biologists separate vertebrates from invertebrates. Chemists distinguish elements from compounds. Rational processing also uses universal rules to solve problems. If two things are both equal to a third thing, then they are always equal to each other. If two people ask for help, and you can only help one, rational principles dictate that the one who is most deserving or has the greatest need should be given assistance.

Top-Down	Bottom-Up
Conscious	Unconscious
Flexible	Resistant to change
Controlled	Automatic
High effort	Effortless
Slow	Fast
Low capacity	High capacity
Evolutionarily new	Evolutionarily old
Uniquely human	Shared with animals
Linked to language	Nonverbal
Abstract	Concrete
Rational	Emotional

Characteristics of conscious, rational, top-down processing versus unconscious, experiential, bottom-up processing.[8]

The top-down conscious mind uses analysis, taking things apart to make them understandable. The bottom-up unconscious mind looks at things as a whole. ***(Source: Left, Professor Otto Wilhelm Thome,*** **Pinus sylvestris** ***L, 1885; right, PIXNIO.)***

Experiential, bottom-up processing is oriented toward specific instances. Each thing is unique. Rather than isolating specific details, as categorization requires, experiential processing takes a holistic approach, viewing things as individuals (a word that originally meant "indivisible"). Instead of using logic to make decisions, it uses feelings. Things are connected not via a conceptual understanding but a feeling of similarity. In this schema, fireflies are like tiny stars, and a ladder represents the ascending path of spiritual growth. If two people ask for help, and you can only help one, your instinct is to help the one you feel closer to.

Some people tend to rely more on rational processing and others on experiential, but it's not an either-or distinction. The two operate independently, so a person can have high levels, or low levels, of both. And how much a person utilizes the two modes of processing can be measured using questionnaires developed by researchers.

People who score high on rational processing tend to be conscientious and believe they can exert control over the world. Those with high scores on experiential processing are intuitive. They rely on gut feelings and are likely to be agreeable and emotionally expressive. They tend to be more trusting, tolerant, and spontaneous, and also enjoy greater than average life satisfaction. People with "complementary processing"—those who score high on both—have better relationships, perceive more meaning in the world, and have a more positive view of themselves and others.[9]

In modern times, we place a high value on rational thought. We measure the strength of rational processing with IQ tests, and they do a pretty good job of not only predicting success in school, but also how much money a person will earn over their lifetime.[10] But there's more to life than material wealth. There are times when it's best to simply be, to follow the example of the poet who "look'd up in perfect silence at the stars." It may not get us a new car or the latest cell phone, but it can enrich our lives in ways that possessions cannot. As we'll see, it's the gateway to the supernatural.

BOTTOM-UP PROCESSING AND THE SUPERNATURAL

As we look more closely at the consequences of the two processing styles, the connection between the unconscious and magic begins to emerge. Experiential and complementary processing are associated with supernatural beliefs and supernatural experiences, including superstitions and paranormal encounters. You might think that people with high scores on rational processing would be less likely to hold supernatural beliefs, but that's not what the data show. Such a high-scoring person is neither more nor less likely to have supernatural beliefs. Only people with low experiential scores (something not necessarily linked to high rational scores) have a lower likelihood of having these beliefs.[11]

Whereas top-down processing takes things apart using analysis, bottom-up processing does the opposite. It connects things based on similarity and other forms of association. This leads to a form of thinking called *sympathetic magic*, first described by Sir James Frazer in *The Golden Bough* book series, his study of comparative mythology and religion.[12] There are two components of sympathetic magic. The first is the law of similarity, which can be summarized as "like produces like." The law of similarity explains why some people believe sticking pins in a voodoo doll will harm the person it resembles, and how a diamond, which sparkles with beauty and lasts forever, represents the promise of eternal love. The Harry Potter series uses many examples of the law of similarity. In one scene, the children brew a potion that allows them to assume the appearance of another person. A key ingredient is a strand of hair from the individual being impersonated.

Researchers studying sympathetic magic found that some volunteers had difficulty throwing darts at a target that had a picture of a baby on it. Because the target looked like a baby, the unconscious treated it like a real baby and spoiled the volunteers' aim. Those with higher experiential processing scores—who are therefore more likely to be influenced by the laws of sympathetic magic—had more difficulty hitting the target than those with lower scores.[13]

The second component of sympathetic magic is the law of contagion. It's the belief that physical contact between two things results in the transfer of some essence or quality. The law of contagion is the reason we value things that have been touched by celebrities. The baseball bat that Babe Ruth used to hit his five hundredth home run sold for more than $1 million.[14] That's a lot of money, and the only thing the buyer got in return for this small fortune (besides a piece of wood) was the magical feeling the bat emanates. If the person who bought it showed it to you, you would probably experience an urge to touch it. It's worth considering why.

The law of contagion was also at work in the behavior of a young woman who received a few grains of stolen moon dust as a love offering from her astronomer boyfriend. She ate them.[15] Sympathetic magic is often seen in fairy tales. In some of the earlier tellings of the Snow White fairy tale, the evil stepmother demands that Snow White's liver and lungs be served to her for dinner. By eating them, she imagines she will acquire Snow White's quality of youthful beauty via the law of contagion.

The law of contagion not only makes things attractive. It can also make them repulsive. Research volunteers were introduced to a man they were told had stepped in dog excrement but who had thoroughly cleaned his shoes. Later, when it was time to sit down, researchers could estimate the volunteers' experiential processing scores by how far away from him they sat.[16]

The law of contagion can also work backwards. People are reluctant to allow personal items to fall into the hands of those they dislike. You wouldn't like to see your enemy wearing the jacket you dropped off at Goodwill.

Earlier, we saw how high stakes and uncertainty can lead to superstitious behavior among elite athletes. An understanding of experiential processing reveals another explanation for this tendency toward magical thinking. Athletes rely on experiential processing more than most people. The reason a tennis player practices her swing over and over, thousands of times, is because she needs to get the mechanics of the movement out of her head and into her muscles. That's another way of saying she needs to move her muscle coordination from top-down rational processing to bottom-up experiential processing, and this takes thousands of repetitions. Although experiential processing is fast, automatic, and effortless, it's resistant to change. It's not flexible like rational processing. The only way to modify bottom-up processing is repetition—that is, experience.

If you've ever had a tennis lesson, you might have been told how to stand when you hit the ball, how to bring the racquet back, and how to perform the forward swing and follow-through. If you try to hit the ball while carrying out these instructions, nothing good will happen. You won't be successful until you've done it so many times you no longer need to think about it. At that point, your conscious attention is no longer directed to your body; it's focused on the ball, your opponent, and the progress of the game. The same is true when you're catching a baseball. If you focus on moving your arm properly, you might get hit in the face. But if you focus on the ball, and let your muscles take care of themselves, you're much more likely to catch it. Conscious interference disrupts unconscious processes—those that are fast, automatic, and much better equipped to do the job.

People whose primary focus is on their physical performance get in the habit of relying on experiential processing, and this habit predisposes them to interpret events from a supernatural perspective. Like athletes, professional musicians must also spend many hours of practice training their experiential systems, and they're likewise more superstitious than most.[17] As athletes and musicians improve their skills, as they move from amateur to professional to elite, they develop an increasingly close relationship with their bottom-up, unconscious circuits.

Athletes and musicians tend to be superstitious because they spend so much time emphasizing experiential processing. We see a similar increase in magical thinking whenever we suppress rational processing because, in both cases, the ratio of experiential to rational processing goes up. Here's how it was done in one experiment. Since top-down thinking is slow, you can bring out experiential processing by making people respond quickly. When volunteers were asked to rate how "literally true" magical statements were, such as "Trees can sense the wind," the less time they were given to answer, the truer the statement was perceived to be.[18] Magical thinking is more common than most people realize. It lies just out of sight of conscious awareness, poised to break out when the conditions are right.

INTUITION AND MAGIC

Since experiential processing takes place outside conscious awareness, the conclusions reached by this form of thinking seem to come out of nowhere. That's because the unconscious mind picks up patterns the ego doesn't see and uses those patterns to form beliefs. These beliefs are what we call intuition, often described as knowing without knowing how we know.[19] Intuition sometimes feels magical because the cognitive sophistication required to develop an intuition about something is beyond the capability of the ego. Science fiction author Arthur C. Clarke writes, "Any sufficiently advanced technology is indistinguishable from magic."[20] We might say the same thing about advanced mental activity.

Not everyone has the same intuitive abilities. Some people are more intuitive than others, and if we could measure the accuracy of people's intuition, we would have another way to determine the strength of their experiential processing system. The studies described so far all relied on self-report questionnaires to determine their volunteers' thinking styles. Well-designed questionnaires produce reliable data, but it's good to supplement them with objective measurements. Objectively measuring something mysterious like intuition sounds tricky, but if we define intuition as pattern recognition that takes place outside conscious awareness, it becomes easier to design tests to see how well different people perform.

One test of intuition involves a circle that can appear in one of four locations on a computer screen. The person taking the test is instructed to press a key depending on where the circle appears, and the computer records how much time it takes to press the correct key. If the location of the circle follows a simple pattern that the test taker succeeds in recognizing, let's say *1-3-4-2, 1-3-4-2, 1-3-4-2* . . . , after a few repetitions she will know in advance which location is next. Then she can press the correct key faster than if the pattern were random. If, on the other hand, she doesn't pick up the pattern because it's too complicated, then we would expect the timing of her key presses to be no different than when the location is random. But that's not what happens. When the on-screen location followed a pattern that couldn't be recognized by the ego, volunteers still pressed the correct key faster than when the circle location was random.[21] They intuitively sensed where it would appear next.

The researchers wanted to make sure the repeating pattern they used was complex enough to overwhelm the capabilities of the ego, so they did some pretests. First, they asked volunteers to take the test without giving them any information about the presence or absence of repeating patterns. Then they repeated the test, but this time revealed to the volunteers that a pattern would be present. When the volunteers used their conscious, top-down capabilities to look for this pattern, their performance got worse. Like the task of choosing the right apartment in Amsterdam, the researchers had successfully created a problem that was best solved with intuition.

The researchers alternated sequences containing complex patterns with sequences that were random, and then compared the pressing time of the two conditions. Some volunteers showed clear differences between the two conditions; they pressed the correct key significantly faster when a complex pattern was present compared to when it was simply random. With others, there was less of a difference, indicating they weren't as intuitive (i.e., their experiential processing systems weren't as good at picking up hidden patterns). Now the researchers had an objective measure of intuition and could evaluate whether it was associated with supernatural beliefs. They specifically looked at belief in divine intervention. Consistent with the self-report questionnaire studies, they found that volunteers who were more intuitive held stronger beliefs in a deity who orders the world in a purposeful way and intervenes to carry out his will.

The study was done in Washington, DC, and to make sure the outcome wasn't a result of cultural factors, the scientists repeated it in Kabul, Afghanistan. They got the same results. Individuals who performed better on the intuition test held stronger beliefs in an intervening/ordering deity. Across cultures, people who had close relationships with their unconscious—indicated by the strength of their intuition—were more likely to understand the world in a way that transcended the materialistic explanations of rational science. They were more strongly influenced by the magical perspective of the unconscious.

A DIFFERENT APPROACH

Neuroscientific and psychological experiments are most successful when they focus on simple behaviors. The experimental method relies on identifying a single factor to be studied and then isolating it from other possible determinants of behavior. This isolation provides confidence that the factor being studied is, in fact, responsible for the behavior of interest. The disadvantage is that it creates an artificial situation that doesn't fully reflect real life. Furthermore, it's limited when it comes to studying the more complex aspects of the human mind. It's like trying to understand how someone walks by isolating the activity of a single muscle fiber. The fishing game study, for example, depended on a contrived setup in which participants were manipulated with lists of words related to cooperation. The results were revealing, but given the complexity of real life, a great deal was necessarily missing from the experimental environment.

An alternative to controlled experiments is naturalistic observation. Rather than imposing artificial constraints on behavior, the investigator simply describes what he sees. Similar to scientific experiments, naturalistic observation is an empirical approach that uses tangible evidence rather than philosophical speculation. This observational approach is what Jung used to understand the clinical material that arose while working with patients. He found that universal themes emerged, themes that were also present in myths, folklore, and fairy tales. He called these themes *archetypes*, and exploring them can help us better understand the mystical link between magic and the unconscious mind.

THE MAGICAL INSTINCT

Until you make the unconscious conscious, it will direct your life and you will call it fate.

C. G. JUNG

ARCHETYPES AND THE UNCONSCIOUS MIND

Many people are familiar with Freud's model of the unconscious, particularly his description of the instinctual drives of sexual desire and aggression. Jung, who was a student of Freud's, also identified the unconscious as the seat of human instinct, but he believed the instincts found there were more varied and sophisticated. He rejected Freud's model as too limited.

Although instinctual behavior is often assumed to be simplistic, even the instincts of nonhuman animals can be remarkably complex. Ants have instincts that make them farmers. They plant, fertilize, and harvest fungus. They're also ranchers. They raise aphids, which produce a sugary liquid the ants eat. The ants help their aphid herds find food, and they care for the aphids' eggs to protect them from predators. None of these behaviors needs to be learned. The skills are coded in their genes. Other animal instincts are equally complex. Weaver birds fashion elaborately woven nests. Bees form sophisticated colonies and communicate the location of food using dance. Instinct goes well beyond the primitive drives of sex and aggression.

Given the complexity of animal instincts, one would think that human instincts would be even more complex and certainly include more than sexual desire and aggression. In fact, human instincts are so complex that our conscious minds are incapable of grasping them. Jung called these instincts *archetypes*, and since they form a part of our shared evolutionary heritage, he described them as "the collective unconscious." Jung believed that just as all human bodies share a common anatomy, human minds share a common psychological foundation that transcends differences in culture and values.[1]

Archetypes are blueprints for big ideas—ideas so big that the conscious mind can only conceive of small pieces of them at any one time. These ideas are fundamental to humanity and concern the most important aspects of human existence. They include characters, such as the hero, the great mother, the naive youth, and the wise old man. They include events, like birth, death, initiation, transformation, and the journey. And they include concepts, such as light, dark, spirit, matter, love, and strife. Archetypes are the bones that give structure to the stories we tell ourselves to make sense of our experiences. Stories built from archetypes are fundamental to our ability to understand the world.

We saw in an earlier chapter how, as we go through our daily lives, we make the events going on around us into a narrative so that we can comprehend what those events mean. It's hard to understand motivations, possibilities, and all the other aspects of a situation, unless we can fit them into a story. This is why the Yerkes primate researchers couldn't interpret their data when they stopped using anthropomorphic descriptions; they lacked an intelligible narrative.

When we can't find a story that fits, we say, "None of this makes any sense." We can't simply accept events as they occur; they have to mean something. It's not enough to know what a person did; we need to know why, because the motivation is essential to the story. When we hear about a conflict, we need to know who the good guy is. We must distinguish the hero from the villain. When we watch a sporting event, choosing a team to root for makes the game easier to follow.

Narrative is a popular word in politics. When ambiguous events occur that could be variously interpreted to the advantage of competing parties, each party rushes to create a narrative, a story that supports their interpretation of the facts. If a party can get its narrative adopted, the public will understand the events the way the partisans intend, because the story determines the meaning. A description of a robbery told from one perspective might support a law-and-order agenda, drawing on the archetype of the predator. Told from a different perspective, it could support a poverty-relief initiative, emphasizing the archetype of hardship. A good narrative is more effective than a good argument. As Daniel Kahneman observes, "The confidence that individuals have in their beliefs depends mostly on the quality of the story they can tell about what they see."[2]

We make mistakes when the story we use to understand a series of events doesn't line up with reality, or when the story has been detached from reality by intentional manipulation, as when advertisers present us with heartwarming scenes to stimulate our desire for sugary sodas or the newest SUV. But despite the risk of making mistakes, we shouldn't suppress our storytelling instinct. If we tried to improve the quality of our judgments in this way, we'd be at constant war with ourselves. We'd become like a harsh parent, wearying ourselves with a never-ending stream of criticism. Our brains evolved to understand events through stories, so the most useful thing we can do is to get better at it. We do this by cultivating a familiarity with the archetypes that structure the stories we tell ourselves. These prototypical outlines of character types and events shape our opinions and beliefs at least as much as objective facts do, and learning them is as important as knowing the principles of natural science that govern the material world.

Archetypes may help us transform information into meaning, but they're difficult for the ego to understand because they embody an alien way of thinking. Earlier we looked at analysis as the quintessential form of conscious thought—separating one thing from another in order to understand them. Analysis is so central to conscious understanding that awareness of our own existence depends on recognizing that we're separate from the things we see around us. We experience ourselves as individuals, not as part of a mass of undifferentiated life or the mystical "all is One." Symbolically, conscious thought has been represented by a sword because it cuts and separates into pieces.

The archetypes are the opposite. They combine—unify—as they represent the essential being of all the things on which they bestow meaning. For instance, each journey is unique, but the journey archetype represents the underlying form they all share. Light can come from candles, stars, and lightbulbs, but light from any source is encompassed by the archetypal idea of light. The unconscious can sometimes take this tendency toward unification so far that it makes no sense to the ego. Opposites are as far apart as you can get from the ego's point of view, but to the unconscious, opposites are united. Archetypes always have a light side and a dark side.

Magic and non-ordinary states of consciousness occur when these alien, irrational archetypes invade the conscious mind. When objects in our environment become associated with the archetypes, they're charged with unconscious energy, and this endows them with mystical power. The ball filled with his father's ashes that John Hinkle Jr. used to bowl a perfect game was invested with the archetypes of love, death, and fatherhood, making it an object of strong psychological influence.

The experience of an archetypal invasion is a numinous one. The numinosity can be so overwhelming that it changes the course of a person's life, or it can be just a glimmer that takes us beyond the mundane routine of the ordinary. If you have children, you may have had a deeply moving experience at their births. Consider the case of a woman who had always defined herself based on her career but experienced a radical change when she held her infant child for the first time. For her entire life, she had focused on work, modeling her life on her mother and mentor, both of whom were successful entrepreneurs. She had achieved early recognition in her field, and felt most alive when she was working. It was her passion. But at the moment of her first child's birth, something unexpected happened. In a flash, taking care of this helpless creature became her new passion. For the next few years, her work was relegated to second place.

One of the tools her unconscious used to bring about this change was the secretion of oxytocin, which rises to four times its normal level during childbirth.[3] This bonding hormone increases the odds that a mother's genes will be successfully passed along to the next generation. But although the brain is responsible for regulating physiological processes within the body, it takes place outside conscious awareness. Agents within the unconscious utilize these processes to influence behavior and realize their goals.

Even when the moment of birth doesn't have life-changing effects, there is always something magical about it. The birth may be nothing more than a pregnant hamster having pups, but it resonates with the birth archetype and becomes a special moment.

We see something similar with the journey archetype. One of the most important journeys we take is when we leave home for the first time. It's a rite of passage in which we transition from the dependency of childhood to the responsibilities of becoming an adult. It's a difficult transition to make, and serious mental illnesses, such as schizophrenia, often debut in the context of this sudden leap down the road toward maturity. But whether the road is bumpy or smooth, the archetype ensures that the emerging adult will return from this journey a changed person, with a new perspective on life and a new self-identity. Like birth, the journey archetype can also give us glimmers of numinosity. A two-hour drive is an unpleasant chore, but setting out on a road trip is an adventure.

THE MOTHER ARCHETYPE

We can get a better understanding of the unconscious perspective—how it conceptualizes the world—by examining one of the archetypes in more detail. The mother archetype is a good choice, since it's arguably the most important of all the archetypes. Think of all the things the concept of *mother* encompasses: fertility, nourishment, protection, tranquility, guidance, an embrace, encouragement, love, to name only a few.

It's important to note that these conscious ideas are not aspects of the archetype *itself* because the archetype is *un*conscious. Rather, they are examples of concepts and experiences that fall under the influence of the mother archetype. They are helpful to our understanding of the archetype the way iron filings sprinkled around a magnet aid our understanding of its magnetic field. The filings aren't the field; they're influenced by the field. Seeing the way they arrange themselves reveals the field's shape and strength. It's the

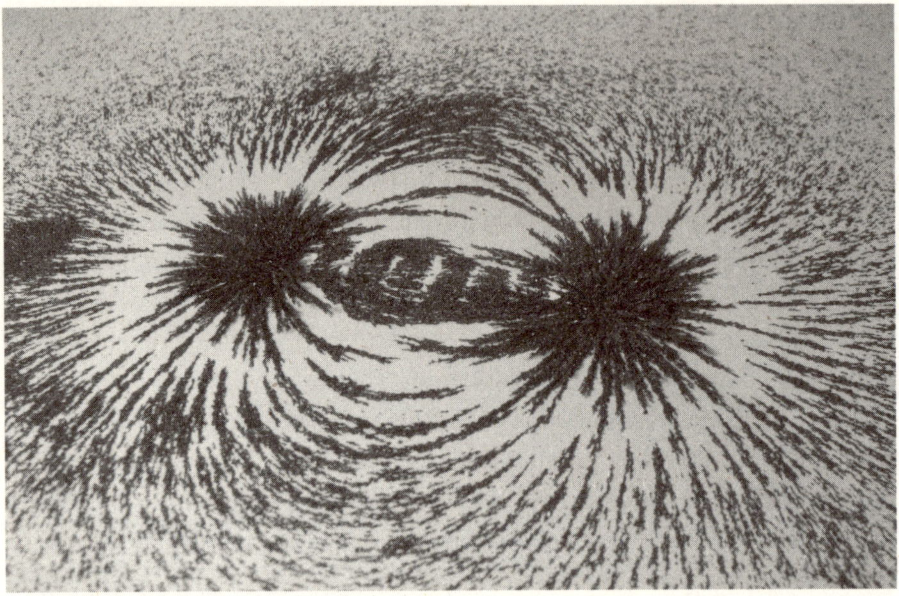

Iron filings are influenced by the invisible magnetic field, but they aren't the field itself. Similarly, conscious ideas and experiences may be influenced by, and help us better understand, unconscious archetypes. *(Source: Windell Oskay via Flickr.)*

same with the conscious ideas and experiences associated with an archetype. By studying them, we can gain a better sense of the archetype, even though the archetype itself cannot be "seen"—cannot become fully conscious.

The influence of the mother archetype on our lives can be symbolized by anything that surrounds, wraps, shelters, and preserves: a valley, the depths of the sea, a forest, a house, or a town. The archetype includes Mother Nature, the womb that produces all life. But in nature, there can be no life without death. Whether an animal is a carnivore or an herbivore, it must kill in order to live. Death nourishes life, and so the fertile earth where life is born is also the grave where life ends. Cave and coffin are symbols of this return to the source of life and are also symbols of the mother archetype.

Life begins in the mother's womb, a paradise where all our needs are fulfilled. The mother and child are one. Birth is the abrupt end of that perfect state of unity. Death is the return. It's our final dissolution, in which our remains are restored to the earth and, once again, we become one with our origin. The archetypal mother is the vessel of life, but she is also the tomb and the darkness that engulfs.

In *The Origins and History of Consciousness*, Jungian psychologist Erich Neumann writes, "As the good mother, [the archetype] is fullness and abundance; the dispenser of life and happiness, the nutrient earth, the cornucopia of the fruitful womb. She is mankind's instinctive experience of the world's depth and beauty, of the goodness and graciousness of mother nature who daily fulfills the promise of redemption and resurrection, of new life and new birth. As the evil mother, [she is] the bloodstained goddess of death, plague, famine, flood, and the force of instinct."[4]

The instinctual unconscious came first and gave birth to consciousness; consequently, the mother archetype can represent the unconscious itself. From this perspective, the relationship between the unconscious and ego is like the relationship between mother and child. The ego is dependent on the unconscious the way a small child is dependent on her mother.

Between the ages of one and two, a child develops the desire to achieve independence. She begins moving away from the safety of physical contact with her primary caretaker (typically the mother) in order to explore the world. During her explorations, the child frequently returns to reestablish contact and emotionally refuel before venturing out again, using the caretaker as a secure base.[5] The conscious mind does something similar. At times it demonstrates independence, but it can't exist on its own for very long. Just as a toddler goes scurrying back to her mother as an emotional sanctuary, the ego returns to the unconscious when it needs to refuel.

A major refueling occurs at night when we sink into the unconscious state of sleep. Even if we've been sedentary all day and have little need for physical rest, we need to sleep in order to find psychological refreshment and renewal. Throughout the day, we periodically sink into varying degrees of unconsciousness in order to avoid running out of conscious energy. Sometimes we're alert and wide awake—fully conscious. At other times we may fall into a reverie and allow our minds to wander aimlessly. After a period of intense concentration, we seek out "mindless" activities, like watching TV, to give consciousness a rest. We may submerge ourselves in the embrace of partial unconsciousness via a warm bath, a lover's arms, or an intoxicating drug.

Death is the ultimate return to the primordial mother of unconscious nature. Neumann illustrates the seductive character of this return by quoting a passage from D. H. Lawrence's poem "The Ship of Death":

The boat dissolves like pearl
as the soul at last slips perfect into the goal, the core
of sheer oblivion and of utter peace,
the womb of silence in the living night.

The most important symbol of the primordial mother is the ouroboros, a snake with its tail in its mouth. It represents unconscious nature, which feeds on itself in a cyclical manner: consuming itself and producing itself from itself, undergoing perpetual destruction and renewal. Life and death are the same to the ouroboric mother. Death nourishes life until life becomes death, one of many cycles that return back upon themselves: the rising and setting of the sun, the seasons of the year, the sowing of seeds and the reaping of the harvest, the circle of life. An ancient prayer to Selene, the Greek goddess of the moon, captures her ouroboric nature: "Thou art beginning and end, only thou alone rulest over all things, everything has its source in thee, and to thee everything returns in the end."[6]

The symbol of the ouroboros can be found in many cultures. The oldest dates back to the fourteenth century BCE and was found on one of the shrines enclosing the sarcophagus of Tutankhamun.[7] In Norse mythology, the ouroboros appears as Jörmungandr, a serpent that encircles the world and grasps its tail in its teeth. The Zohar, one of the books of the Jewish Kabbalah, describes the enormous serpent Leviathan as having its tail in its mouth. Hindu religious art shows the ouroboros forming part of the base on which the earth rests, and according to the Yoga-kundalini Upanishad, a serpent with its tail in its mouth lies half asleep at the base of the body and represents divine power. The widespread distribution of the ouroboros—its emergence in symbolic form around the world and throughout the ages—suggests that it may represent a fundamental aspect of the human psyche.

The ouroboros reflects the creative power of the unconscious. Neumann writes, "The self-generating power of the soul is man's true and final secret . . . These images, ideas, values, and potentialities of the treasure

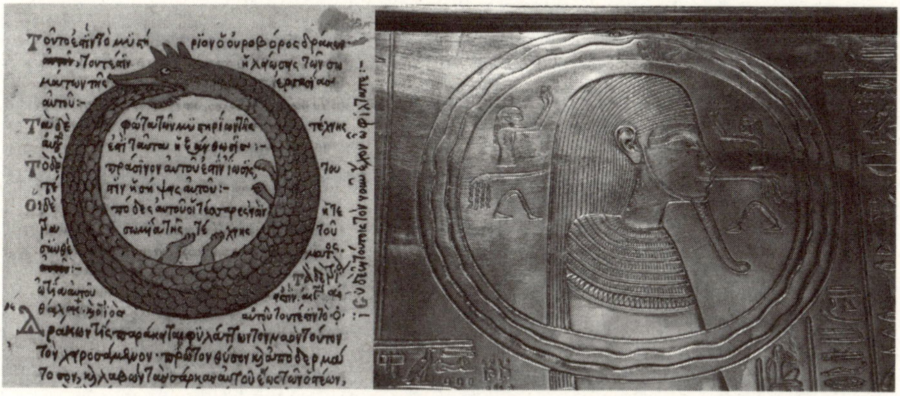

Left: An ouroboros from a fifteenth-century alchemical manuscript (1478). Right: The oldest known ouroboros, from the fourteenth century BCE, carved on a replica of Tutankhamun's gilded shrine. *(Sources: Left, Wikimedia Commons; right, Carsten Frenzl via Wikimedia Commons.)*

hidden in the unconscious are brought to birth and realized by the hero in his various guises."[8] The fertile mother not only gives us a physical existence at our birth; she remains with us throughout life, deep within the psyche, perpetually renewing our vitality.

A CLOSER LOOK AT NOURISHMENT

So far we've only scratched the surface of the mother archetype. Millions of stories have been written that touch on the theme of motherhood, and each one illustrates a different facet of the archetype. It would be impossible to adequately describe what the idea of motherhood represents within the human psyche; it's too big for us to analyze. But let's see what we can do with just one piece, nourishment. Let's look at how that one piece of the mother archetype influences the workings of the human mind and generates numinous experiences.

Nourishment begins at birth with the infant suckling at his mother's breast. Human infants are helpless at birth, and so the relationship with the mother is one of life and death. The symbolic importance of suckling is reflected in mythology. Heroes and gods are sometimes portrayed as being nursed by animals, symbolizing their acquisition of bestial power that allows them to perform superhuman deeds (an example of the law of contagion). The legendary founders of Rome, Romulus and Remus, were nursed by a wolf. Zeus, king of the Olympian gods, was nursed by a goat. The Earth itself is sometimes portrayed as a mother who nurses her children. The alchemical text *Atalanta Fugiens* contains an illustration of the Earth nursing an infant, flanked by Zeus on her right and Romulus and Remus on her left.

As the child grows, his relationship with his mother and the way she provides nourishment evolves. Instead of nursing, she prepares meals for him. Cooking for someone can be an act of love, especially when it is a mother

Mother Earth from the alchemical treatise *Atalanta Fugiens* (1617). *(Source: Wikimedia Commons.)*

preparing a favorite dish for her child. When an adult returns home for a visit, being served a favorite food from childhood links the adult to the past. The memories evoked by the tastes and smells can have a powerful, sometimes even numinous, impact.

Perhaps the most famous description of a numinous experience involving food can be found in Marcel Proust's *In Search of Lost Time.* As an adult, he returns to the town where he grew up. His mother offers him a cup of tea and a small cake called a madeleine, a snack he used to enjoy as a child. The taste fills him with joy and a sense of immortality. The commonplace fades away as the eternal erupts from his unconscious, culminating in a feeling of supernatural possession by the immortal "essence" that is overwhelming him:

> I raised to my lips a spoonful of the tea in which I had soaked a morsel of the cake. No sooner had the warm liquid mixed with the crumbs touched my palate than a shudder ran through me and I stopped, intent upon the extraordinary thing that was happening to me. An exquisite pleasure had invaded my senses, something isolated, detached, with no suggestion of its origin. And at once the vicissitudes of life had become indifferent to me, its disasters innocuous, its brevity illusory—this new sensation having had on me the effect which love has of filling me with a precious essence; or rather this essence was not in me, it *was* me. I had ceased now to feel mediocre, contingent, mortal. Whence could it have come to me, this all-powerful joy?[9]

Love and care are positive aspects of the nourishment archetype, but there's also a dark side, reminding us that early humans were not only predators, but also prey. Being eaten is one of our greatest fears. In medieval paintings of Hell, lost souls are often portrayed as being swallowed and eaten. We speak of "death's maw," a "consuming disease," and being "eaten up with jealousy." Poison is another manifestation of the dark side of the nourishment archetype, one that plays a prominent role in myths and fairy tales. There, poison is often hidden within some desirable food or drink, reminding us that the light and the dark are intertwined.

Researchers at the University of Pennsylvania did an experiment in which they evoked the poison archetype to demonstrate its magical effects.[10] The researchers showed the participants a fresh, unopened bag of Domino sugar. Still in plain view of the participants, they opened the bag and poured sugar into two bottles until the bottles were each about one-third full. They then gave each participant two peel-off labels. One label had the words "Table Sugar" and the other label said "Sodium Cyanide." The participants were asked to put one label on each bottle. Next, a spoonful of the white crystals was taken from one of the bottles and added to a glass of water, and the procedure was repeated with the other bottle. Finally, the two glasses were given to the participants, who were asked which liquid they would like to taste. Despite having witnessed the bottles being filled with sugar from the same bag, and despite the fact that they, themselves, had put the labels on the bottles, most of the participants avoided the liquid that contained sugar from the bottle labeled "Sodium Cyanide."

The researchers described what happened as a form of sympathetic magic. The act of attaching a label that says "Sodium Cyanide" to a bottle of sugar psychologically transformed it into poison—even though the participants knew firsthand that it was just sugar. You can try labeling a bottle of water "Sodium Cyanide" yourself and then seeing what it feels like to drink it. Some people will experience a creepy feeling as they drink the water.

The closer one moves to the mother archetype, the more eerie these feelings become. Imagine what it would feel like to sip a cup of warm baby formula. The feelings produced would probably be uncomfortable, but not the same kind of uncomfortable as going to work with a stain on your shirt. Sipping the baby formula might produce a weird, spooky discomfort.

What is baby formula? It's dried milk powder, whey protein, sugar, and other nutrients. Protein shakes have many of the same ingredients, but protein shakes aren't spooky. The difference is the label. The process that turns milk powder and whey protein into baby formula is the same process that turned the sugar into sodium cyanide. Since baby formula is closer to the mother archetype than poison, it packs a greater emotional punch.

As with the mother archetype more generally, there are countless other aspects of the nourishment archetype—so many that it would be impossible to fully understand the depths of this idea, impossible even to give an approximation of it. That's the nature of archetypes. They are fundamental to human nature and human experience—they shape the way we understand our lives—but we can't comprehend them because they're too big, too potent, and too powerful.

GENETICS

The nebulous nature of archetypes makes them difficult to understand, so it will be helpful to briefly switch gears and look at them from the modern scientific perspective. Specifically, we need to ask: What is an archetype, really? Can we hold it in our hands? Can we understand it within a materialist model?

We've described archetypes as a form of instinct, so we can shed light on this question by asking where instinct comes from in nonhuman animals. The answer is clear: it comes from the DNA found in the nucleus of the organisms' cells. As you may recall from biology class, DNA is a code, like the ones and zeroes of computer code, or the twenty-six letters of the alphabet. DNA has four "letters" called bases: cytosine, guanine, adenine, and thymine, abbreviated as C, G, A, and T.

It's not hard to imagine genes containing the information needed to build a pair of lungs, but it's more difficult to imagine genes encoding the instinct that gives the weaverbird the skill to build its nest. It's even harder to imagine combinations of C, G, A, and T coding for the unfathomable depths of the mother archetype.

In some ways, it's the insubstantial nature of archetypes that makes this encoding so hard to imagine. The best explanation we have is that archetypes emerge from the ways in which brain cells organize themselves into circuits. Despite the vagueness of this description, and how little it seems to explain, the ability to inherit patterns of thought and behavior is supported by numerous studies of identical twins.

Identical twins have the same DNA. If they're separated at birth and raised in different families, we can conclude that the similarities they share are a result of their genes rather than the environment they grew up in: nature, not nurture. We wouldn't be surprised if they had the same eye color or the same hair color. But would you expect them to drink the same brand of beer or pursue the same career?

One of the most remarkable cases of twins separated at birth is the Jim twins. Here's how the twins were described in a *National Geographic* article:[11]

> Born in Piqua, Ohio, in 1939, Jim Springer and Jim Lewis were put up for adoption as babies and raised by different couples, who happened to give them the same first name. When Jim Springer reconnected with his brother at age 39 in 1979, they uncovered a string of other similarities and coincidences. Both men were six feet tall and weighed 180 pounds. Growing up, they'd both had dogs named Toy and taken family vacations in St. Pete Beach in Florida. As young men, they'd both married women named Linda, and then divorced them. Their second wives were both named Betty. They named their sons James Alan and James Allan. They'd both served as part-time sheriffs, enjoyed home carpentry projects, suffered severe headaches, smoked Salem cigarettes, and drank Miller Lite beer. Although they wore their hair differently—Jim Springer had bangs, while Jim Lewis combed his hair straight back—they had the same crooked smile, their voices were indistinguishable, and they both admitted to leaving love notes around the house for their wives.

One might wonder whether this was simply a one-in-a-million coincidence, or they were just adopted by families that were very alike. But other twins have similar stories. Paula Bernstein and Elyse Schein were part of a secret research project in the 1960s and 1970s that separated identical twins as infants and then followed them over time to better understand how genes and environment contribute to human development. When Paula and Elyse were reunited at age thirty-five, they discovered that they shared the same taste in books and music, had both been editors of their high school newspapers, studied film in college, and had careers as writers.[12] If C, G, A, and T can code for music, beer, and career preferences, maybe they can also code for the unknowable archetypes.

While it's useful to explore mental phenomena from a biological perspective, that perspective has its limitations. Understanding archetypes as

DNA codes that generate characteristic patterns of brain activity is called a *reductionist* approach, because it reduces the human experience to molecular interactions. It strips away the subjective component—what the experience feels like. This approach has its uses, but it's important to remember that it's subjective feelings that determine how we experience ourselves and the world. They have the greatest impact on the choices we make and on our fate in life.

When a neuroscientist describes electrochemical signals flowing from one part of the brain to another, it's tempting to yawn and say, "Big deal." The process has been reduced to something almost trivial. But the subjective experience can change a person's life. A man who falls in love feels reborn. A religious convert may abandon everything she used to value to embrace a life of poverty and service. A soldier invested with the archetypal spirit of courage may sprint across a battlefield to rescue a friend, oblivious to the machine gun fire all around him. An inventor may be struck with inspiration that changes the world, and an artist may produce a work of awe-inspiring beauty. Encountering the numinous can be ecstatic and life altering. Explaining the human experience of the numinous by describing the behavior of brain circuits is like trying to explain the beauty of Monet's *Water Lilies* by making a statistical chart of the pigment distribution across the canvas. It's an interesting thing to do and may be useful in certain contexts, but it has little to do with why the painting moves so many people.

Although archetypes are humanity's shared inheritance, they express themselves differently in each individual. Archetypes structure our thoughts the way a blueprint structures a building. A blueprint embodies the idea of a building, but the actual reality depends on many other factors, such as the skills of the construction workers and the building materials they use. The blueprint of an archetype must be filled in by an individual's life experiences, making each person's experience of the archetype unique.

Genetics may allow us to understand archetypes within the materialist model, but it's a model that gives us only a vague idea of how the brain generates human experiences. To understand the archetypes on a deeper level, we need a different strategy, one that shifts our perspective from the objective to the subjective, and from the mundane to the divine.

SYMBOLS EXPRESS WHAT THOUGHTS CANNOT THINK

Archetypes generally stay in the background, shaping the way we understand the world, but occasionally they erupt into consciousness. Fleeting encounters with archetypes occur in dreams and during numinous experiences. But the most enduring form of archetypal eruption is found in art, literature, and other forms of aesthetic expression: it's the *symbol.*

Symbols are bridges that cross the chasm between the conscious and unconscious worlds. They are things we can perceive with our senses that connect us to the invisible archetypes. They can be natural objects, such as the sun, moon, and trees, or they can be produced objects, such as a sword or a shield. Earlier, we noted some of the symbols of the mother archetype, which include the earth, the ocean, and containers, such as cups and bowls.

To better understand symbols, it's useful to compare them to what linguistics refers to as *signs*, which are quite different. A sign is a unit of communication in which there is a clear one-to-one connection between a material thing and what that thing signifies. For example, the combination of the letters o-r-a-n-g-e signifies the fruit. When you read the word, you know exactly what it signifies, neither more nor less. The same is true of the golden arches, which signify the McDonald's restaurant chain. When you see them, you know exactly what they mean. A sign gets its meaning from the thing it signifies, and that thing is knowable. Symbols, by contrast, get their meaning from archetypes, which are unknowable. Just as archetypes are too big for the ego to comprehend, so are the meanings of symbols. Symbols express what thoughts cannot think.

C. G. Jung, who wrote extensively about how symbols connect the conscious mind to the archetypes, first began to investigate their role in psychology after encountering them in the dreams of his patients. He was startled to see symbols that are commonly found in mythology and art in the dreams of patients who had no familiarity with these fields of study. He concluded that symbols spontaneously arise in dreams, visions, and ecstatic religious states. Going further, he noted that recognizable symbols are found among the relics of prehistoric humans, making them as old as

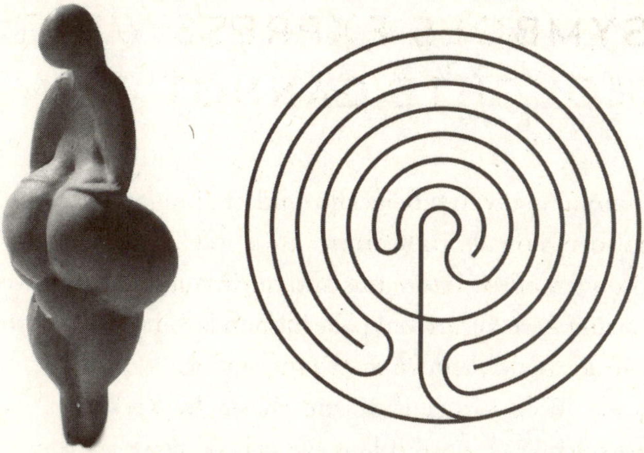

Left: A replica of the Venus of Lespugue, a 25,000-year-old statuette thought to symbolize fertility. Right: The Hopi symbol Tapuat, representing a mother and her child as well as the umbilical cord and the stages of life. *(Source: Left, José-Manuel Benito/Locutus Borg; right, Symbolikon Library.)*

our species, and suggesting that symbol formation is an important biological function.

As noted, the experience of the archetype is influenced by individual psychological factors. That makes each archetypal experience unique, not just among individuals but also within the same individual at different times in their life. The same is true with symbols. Although its theme is universal, the specific meaning of a symbol is different for every person and carries different meanings at different times. The question, "What does [the symbol] mean?" has no answer, because symbols have no fixed meaning. The meaning has to be discovered by each person at each encounter. That's another difference between signs and symbols. Signs mean the same thing to everybody, and their meaning never changes. Anybody familiar with the golden arches knows they signify McDonald's. There's no variation from person to person.

Examining one symbol in detail will throw additional light on how they work in relation to the archetypes. A wedding ring symbolizes the archetype of *mating for life*. This instinct is seen not only in humans but also in other animals, such as swans and prairie voles. Humans, of course, experience infidelity and divorce, but other typically monogamous animals will also sometimes abandon their mate or mate polygamously. When the archetypal blueprint is made real, it degenerates to some extent. The symbol, however, reflects its pure form. As with any other symbol, we can't describe what a wedding ring means, but we can examine some of the associations it embodies that shed light on the human archetype of mating for life.

Wedding rings are typically made of gold. Unlike other metals, gold doesn't rust, so it's a symbol of unchanging eternity. Gold is also associated with royalty, and marriage symbolically transforms the bride and groom into a queen and king. They become rulers of their family, a tiny kingdom. Couples married within the Greek Orthodox Church have an actual "crowning ceremony," in which the priest places *stefana*, or wedding crowns, on their heads. In other traditions, the crown may be replaced by a bridal headpiece.

Wedding rings are round, another symbol of eternity because a circle has no beginning and no end, an echo of the ouroboros. The circle also symbolizes mystical transcendence because it has an infinite number of sides, embodying perfection and unity.

A ring can surround and demarcate a distinct space. It may be a sacred place, a place of safety, or a place where it is lawful for only certain people to enter. It may be a place of transformation, like a fairy ring. But a ring can

Left: Stefana: wedding crowns worn at a Greek Orthodox wedding. Right: A crown-like bridal headpiece. *(Source: Left, Zdeněk Kratochvíl via Wikimedia Commons; right, Faylyne via Flickr.)*

also designate a place of conflict and danger, such as a boxing ring or a trap. A wedding band not only symbolizes connectedness, it can also imprison the wearer, like a shackle. It announces that the wearer is off limits.

It's important to remember that these associations help put the symbol of the wedding ring in context, but they do not tell us what a wedding ring means. If we could understand precisely what the symbol meant, it would no longer be a symbol; it would become a sign. The nature of symbols is that their meaning cannot be consciously comprehended.

Despite the knowable associations just described, wedding rings retain their symbolic power and numinous qualities for most people. They have emotional qualities that can't be adequately described with words. The feeling of wearing a wedding ring—or not wearing it—goes well beyond the sensation of metal on skin, and losing a wedding ring is a kind of catastrophe, a bad omen for the marriage. Such a loss on the part of one partner stimulates strong emotions in both parties.

Thus, a wedding ring is both a sign and a symbol. It's a sign insofar as we can describe some of its meaning in words, and a symbol for the meaning we can't describe. A pure symbol would be something you couldn't describe at all, because its meaning has nothing to do with words. Encountering it—experiencing the nearness of the archetype it symbolizes—would provoke a speechless wonder and touch the deepest depths and highest heights of the soul. Different things act as symbols for different people, but for many, a sunset fits the bill. If you imagine the feeling you experienced watching the most inspiring sunset you ever saw, it will give you an idea of how a symbol works. It sometimes feels like looking into another world, even though the experience is being generated from within the psyche.

While they retain ambiguity in their meaning, symbols have the advantage over archetypes of being perceptible to the conscious mind. By creating a channel into the unconscious, symbols allow the energy of the archetype to flow up into the ego. A wedding ring may make it easier to resist the promiscuity instinct that competes with the mating-for-life instinct. Symbols of patriotism, such as a country's flag, may help ordinary people find the courage to fight as soldiers. Symbols are energy transformers. They make the energy stored in the unconscious available to the ego.

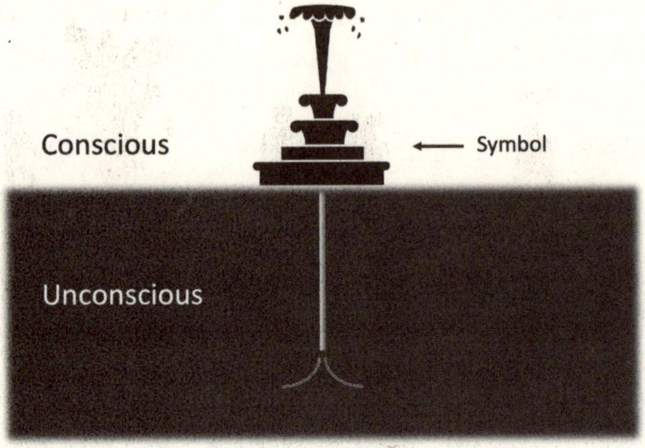

Symbols are a link to the unconscious.

OVERWHELMED BY THE UNCONSCIOUS

When the ego incorporates even a small amount of unconscious material, it leads to enhanced energy, passion, and an increased capacity for work. But if we are able to fully integrate the unconscious into consciousness, their union produces the complete individual, incorporating all aspects of the mind, which Jung named the "self." It's an ambitious goal and probably best thought of as an aspiration or a guide rather than something that can be realistically achieved. Nevertheless, Jung believed that personal growth is characterized by ever-increasing integration of the two parts of the mind. He described a special instinct within the unconscious that urges us toward this integration, which he named the "transcendent function" because it transcends the division that separates the ego from the unconscious.

Swiss Jungian psychologist and scholar Marie-Louise von Franz writes, "When the conscious and unconscious learn to know, respect, and accommodate one another, a person becomes whole, integrated, calm, fertile, and happy."[13] That's the goal. However, it's important to appreciate the dangers that lurk along the path. Remember, just as the ego may try to play the tyrant by denying the role of the unconscious, the unconscious, because it is

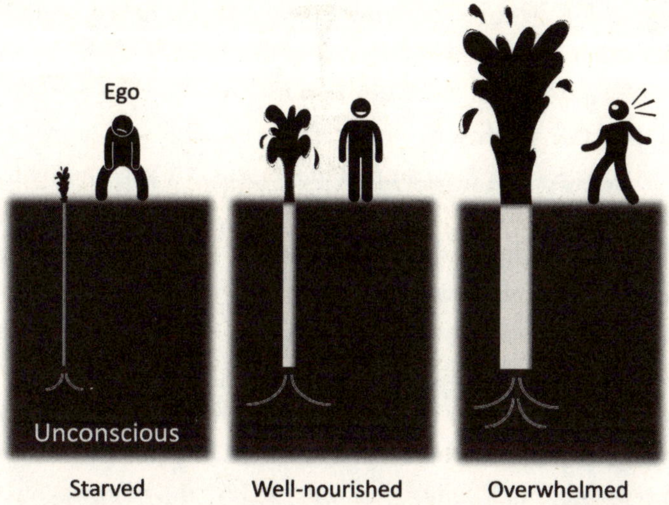

It's important to have just the right amount of intermingling between consciousness and unconsciousness. If there is too little, the ego doesn't have access to the instinctual drives and becomes weak. If there is too much, the ego can't cope with the onslaught.

so much larger, can overwhelm the ego. As Jung writes, "It is as if one were digging an artesian well and ran the risk of stumbling on a volcano."[14]

There's a psychiatric illness called "Florence syndrome" in which people who visit the Italian city become overwhelmed by the sheer number of artistic masterpieces the city is famous for, masterpieces filled with archetypal symbols. The staff at the local hospital are accustomed to treating tourists suffering from dizzy spells and disorientation. "Jerusalem syndrome" is similar, though more severe. Some tourists develop a religiously themed psychosis, which typically resolves after leaving the city.

In stories and myths, being overwhelmed by the unconscious is sometimes symbolized by the hero sinking to the bottom of the sea, being washed away in a flood, or suffering imprisonment in an underground dungeon. Real people who become overwhelmed by their unconscious may go through life in a dreamy state, seeming to live in their own reality. Consider the case of a young man in his twenties who sought treatment for a problem called "maladaptive daydreaming." He was unable to stop himself from fantasizing, no matter what he was doing in real life. This daydreaming made it nearly impossible for him to maintain social relationships, because whenever he was with other people, his mind was somewhere else. He described a trip he took to Norway in which he went on a fjord-climbing expedition. As he clung to the side of a cliff, surrounded by indescribable beauty, he saw none of it. His mind was far away, trapped in a fantasy of pretending to be a rock star on an imaginary stage.

A person overwhelmed by their unconscious may become overly influenced by instincts, losing the ability to consciously reflect on their choices—even losing part of their humanity—as they get too close to the animal side of their nature. They may suffer uncontrolled anger, violence, lust, greed, or desire for power. In myths and fairy tales, we see this problem symbolized by characters who are transformed into animals.

It's also possible that being overwhelmed by the unconscious can make one feel superhuman, as if one were possessed by a god. A person who gives in to rage feels both greater and less than human. The rage fills him with a power that sweeps away obstacles and boundaries. It can be intensely pleasurable to run roughshod over social conventions as one gives free rein to passion. And yet, although he may feel godlike, he knows he's acting by instinct, the way an animal does, rather than thinking rationally, like a man. Within the unconscious one finds a combination of the bestial and the divine, symbolized in Greek mythology by the satyr, a creature who was half goat and half god.

The unconscious is powerful and dangerous. We shouldn't ignore it, but we shouldn't uncritically embrace it or surrender to it, either. The goal is to find balance, to bring unconscious contents into consciousness in a way that doesn't overwhelm the ego. Becoming whole is perhaps the most important thing we do in life. It would be naive to think that it would be easy or free of danger. As von Franz writes, "For wherever there is a pearl there is a monster lying on it, wherever there is a treasure there is a snake a wound around it, and wherever there is the water of life there is a lion guarding it. You cannot get near the self and the meaning of life without being on the razor's edge of falling into greed, into darkness, and into the shadowy aspect of the personality."[15]

INDIVIDUATION

If you can face the instinctual forces of the unconscious without being overwhelmed by them, you have the opportunity to integrate them with the ego to develop the self and become the person you were meant to be. Because the process is what makes a person a unique individual, Jung named it "individuation." As you go through this process, you become less and less like the undifferentiated crowd and more and more like yourself.

Undifferentiated people act and think in similar ways. One way people can lose their individuality is by adhering too faithfully to the collective consciousness. The collective consciousness is made up of the values, attitudes, and opinions that are consciously shared within a group. A harmless example is a teenage fan club. Members dress alike, talk alike, and have similar opinions. Belonging to a homogeneous group can make a person feel safe, and it's common during adolescence, when frightening physiological and psychological changes are taking place and emotional safety is at a premium.

It's not always harmless, though. A darker example is a totalitarian regime. Opinions and beliefs handed down by the authorities are accepted uncritically, and the conscious attitude of the masses is homogenized. Spirituality, intuition, and dreams are all suppressed; those manifestations of human nature upset the regular order and are enemies of the state.

Children's author Madeleine L'Engle warned of the danger of loss of individuality in her Newbery Medal acceptance speech in 1963. She said, "There are forces working in the world as never before in the history of mankind for standardization, for the regimentation of us all, or what I like to call making muffins of us, muffins like every other muffin in the muffin tin."

Lack of differentiation can also come from the opposite direction, from the collective *un*conscious. Pure instinct is undifferentiated, and when instinct alone guides our behavior, we revert to our animal nature. To illustrate how unconscious instinct can flood the psyche and wash away individuality, Jung used the example of ancient Dionysian orgies. Dionysus was the Greek god of wine, fertility, and religious ecstasy. During his festivals, worshipers performed ritualistic dances, got drunk, sacrificed animals, and engaged in a sexual free-for-all. Jung describes humans in this state as "nothing but sheer Nature, unbridled, a raging torrent." He writes, "in the Dionysian state the

Greek was . . . robbed of his individuality, dissolved into his collective components, made one with the collective unconscious (through the surrender of his individual aims), and one with 'the genius of the race, even with Nature herself.'"[16] Despite the loss of individuality (or perhaps because of it), Dionysian orgies were quite popular. The human condition can be a difficult one, often marked by anxiety and uncertainty. It can be tempting to embrace the opportunity to descend into the simplicity of our instinctual nature.

It's only the combination of consciousness with the unconscious that leads to individuality. If a good citizen in a totalitarian regime were to develop an intuition that the things he was told to believe weren't quite right, or had an inspiration—a new idea that might bring about fundamental change—he would become an individual. Conversely, if a devotee of Dionysus began to think rationally during an orgiastic ritual, she would immediately differentiate herself from the frenzied mass.

Intermingling the rational ego with animal instinct isn't always comfortable. Philosophers throughout history have commented on the dual nature of humanity. We are mind and body, spirit and flesh, reason and instinct. In many cases, the first is portrayed as good and the second as bad. Fantasies of disembodied intellects are common. A short story written by American author Kurt Vonnegut titled "Unready to Wear" describes a group of people who discover how to leave their bodies and live an idealized life as incorporeal consciousnesses. The character who first made the discovery, Dr. Ellis Konigswasser, says, "The mind is the only thing about human beings that's worth anything. Why does it have to be tied to a bag of skin, blood, hair, meat, bones, and tubes?" It's tempting to have contempt for our animal side, but it's a mistake. We cannot become ourselves until we come to terms with it and make it our ally.

The unconscious counterbalances the ego. Dreams are one of the most common ways the unconscious communicates with the ego, and von Franz notes that "dreams heal, compensate, confirm, criticize, or complete the conscious attitude of an individual."[17] The unconscious brings balance, which is healthy, but accepting the balancing influence of the unconscious can also be frightening, because it forces the ego to confront ideas and attitudes that are alien to it. A pacifist might dream of aggression, and a fighter might dream of surrender. This doesn't mean that people need to become their opposite.

However, they do need to be reconciled with the opposite side of themselves and possess the ability to call on either aggression or surrender when each becomes the truest expression of who they are.

The tension that exists between the complementary opposites of the ego and the unconscious forms the basis of another important archetype, the enemy brothers. The enemy brothers are two individuals who are intimately connected, but the link is a hostile one. Set and Osiris from Egyptian mythology are ancient examples of this archetype.

In modern storytelling, the enemy brothers are recognizable in buddy films, such as *Lethal Weapon*, *Men in Black*, and *The Heat*. The plot involves two characters who hate each other as the story begins, but around the midpoint realize they can't accomplish their goals without working together. By the end of the movie, they have developed a deep mutual affection. Many romantic comedies follow a similar plot, in which the protagonists start out disliking each other, subsequently fall in love, and ultimately achieve union through sex or marriage, symbolizing the process of individuation.

The purpose of life is to become your true self. Sometimes that requires hard labor, wrestling with obstacles that inevitably block your path. But at other times it simply means looking up "in perfect silence at the stars." If we can hear the unconscious speak, we have an opportunity to take a small piece of it and incorporate it into the ego. If we can find a way to accommodate the hidden forces inside ourselves, we will accomplish far more than the ego can manage on its own. There are unexplored depths within us, depths that hold untold riches.

To explore these depths, we must enter the realm of the irrational. The unconscious thinks in a different way and uses a different language. Years ago, aspiring doctors had to learn Latin before they entered medical school. They had to learn the language of medicine before they could learn to use the tools of medicine. It's the same with individuation. If we want to make progress down that road, if we want to learn to use the tools that will make us the complete person we were meant to be, we have to acquire a new vocabulary.

Learning a new language involves more than just substituting one word for another. There are things that can be expressed in Latin, for example, that can't be fully translated into any other language, because embedded within languages is a particular way of understanding the world. Many Eastern languages express the relationship between the speaker and the listener with more subtlety than Western languages. They have different levels of formality—different words, prefixes, and verb declensions—that reflect the status of the person being addressed. Often, the relationship that the various levels of formality signify cannot be precisely described in a Western language. It's the same with the language of the unconscious. A list of definitions isn't very helpful; in effect, we have to immerse ourselves in the culture.

Learning the language of the unconscious means becoming familiar with the traditions of magic. Stories about magic often don't make logical sense because they use symbolism and metaphor to communicate their ideas. There is never a single meaning; a host of many different meanings is implied. Consequently, the language of metaphor and symbolism doesn't exist on its own; it must be interpreted by each individual who hears it. Rational communication tends to have a single meaning. If someone says, "My house is about two miles from my work," everyone can agree on the information being conveyed. But when we hear a story about two children finding a house made out of candy, we're not quite sure what it means, and each of us will interpret it in our own way. However, as we become more comfortable making these interpretations, the world of our unconscious begins to unfold.

As we begin our exploration of the vocabulary of the unconscious, we'll start near the surface, by looking at a part of the unconscious that lies closest to the well-lit regions of consciousness. There we'll find the shadow, a dark beast that guards the gateway to the deeper regions of the unconscious. Focusing on one's individual experience of the shadow is a relatively accessible way to begin gaining familiarity with the nature of psychological dark matter. Before we look at the darkness in the traditions of magic, we'll first look at it in ourselves.

THE SHADOW

I have a hand mirror and inside the hand mirror is another face . . . I sensed in the dream that it was an evil face and so if I look directly into its eyes and stare it out, I can somehow get rid of it. I can feel that I'm almost getting rid of it but then suddenly I lose concentration and its face breaks and I see my own.

BIL, "ANOTHER FACE," FROM THE INTERNATIONAL ARCHIVE OF DREAMS, 2013

AN EXPERIMENT

In a basement laboratory at Yale University, two research volunteers were led into a room where one of them was strapped to an electric chair. The scientist conducting the experiment explained that this man was the Learner. His job was to memorize a list of words. The straps were necessary to prevent him from escaping. The other volunteer would be the Teacher. His job was to read the list and, if the Learner made a mistake, to administer an electric shock. For each mistake, the strength of the shock increased by 15 volts up to a maximum of 450 volts, enough to kill a person. The Teacher was given a sample shock—not a large one, but enough to give him a taste of the pain he would be inflicting. Then the Teacher was taken to another room where he could hear, but not see, the Learner, and the experiment began.

As the Learner started to make mistakes and the shocks grew in intensity, the sounds of a man screaming in pain filled the room. By the tenth shock, the Learner was begging to be set free. But the Teacher, encouraged by the scientist, continued to administer stronger and stronger electric shocks until, at last, the final shock was given, and there was silence in the room.

It was 1961, three months after Nazi war criminal Adolf Eichmann went on trial, and psychologist Stanley Milgram was researching the nature of evil. Specifically, he wanted to find out to what degree he could induce ordinary people to perform acts of cruelty. The setup was fake. The Learner was an actor, and the electric shocks weren't real. Most of the research volunteers figured it out, but not everyone, and among those who believed they were actually harming the Learner, about a third continued to administer shocks up to the maximum voltage. This was a far greater proportion than predicted. The researchers had underestimated the average person's capacity for evil.

Most people underestimate not just others' capacity for evil, but their own as well. As Robert Simon, an expert on the criminally insane, titled his book, *Bad Men Do What Good Men Dream*. Good people and bad people often have the same impulses. The difference is that some people choose to act on those impulses, while others oppose them.

Primitive urges, such as the desire to inflict pain, are frightening, and at times they're too much for us to accept. So we banish these unacceptable aspects of ourselves to our unconscious, where they lie hidden, out of sight, but ready to burst forth at any moment, sometimes shocking us to the core. Jung called the collection of these unacceptable parts of the self "the shadow."

REPRESSION

Since the shadow is unconscious, we can't see it in ourselves. It's not so difficult, however, to see the shadow in other people, and so we're often surprised when they cannot. Most of us have known mean-spirited people who make those around them miserable while maintaining absolute confidence in their own moral superiority.

The process by which unacceptable thoughts are forced out of consciousness is called *repression*, a concept we encountered in an earlier chapter when we looked at research volunteers who, when they were told to stop thinking about sex, found that the thoughts kept resurfacing despite their best efforts. In that experiment, the volunteers were told to make a conscious effort not to think about the conflicted topic, but people also engage in repression without realizing it. In both cases, the result is the same: the repressed impulses emerge unexpectedly in ways we can't control.

It can be embarrassing when the shadow rears its head. Have you ever blurted out something that you immediately regretted? *Where did that come from?* you might have asked yourself. *That's not like me to say something so insensitive.* Or have you ever accidently broken something and been accused of doing it intentionally? Take the example of a married couple who were experiencing a bumpy patch in their relationship. One day, the wife was pulling out of the garage and smashed into her husband's new sports car. She swore it was an accident, but he was angry, pointing out that, the day before, she had criticized the car as an extravagance.

We're more likely to forget things—repress them—when they make us uncomfortable. When parts of the brain produce thoughts and ideas that are unacceptable to the ego, conflict occurs, causing a state of discomfort that must be resolved. The easiest way to resolve it is to pretend it never happened, to disavow the unacceptable thoughts. The memory is scrubbed clean.

This process was demonstrated in a laboratory by scientists at the University of Bonn.[1] When they caused volunteers to experience internal conflict, the volunteers had difficulty remembering their own reactions. The researchers displayed a variety of sentences on a screen, and then asked the volunteers to say the first three words that came into their minds, a form of free association. Some of the sentences were neutral: "Sometimes my mood is influenced by the weather." Some were negative, but free of conflict: "I get annoyed when I am stuck in a traffic jam and I have an important appointment." And some involved conflict: "I actually only feel good when someone is taking care of me," a statement intended to stimulate the emotionally difficult problem of balancing the desire to be cared for with the desire for self-sufficiency. The volunteers were shown a total of twenty-four sentences—six neutral, six negative, and twelve conflicted—and then were given an unexpected memory recall test. They were shown the sentences again and asked to repeat the words they had associated with the sentences.

Recall was almost exactly the same when it came to neutral and negative sentences; the volunteers forgot about 45 percent of the words they had freely associated. But when it came to conflicted sentences, things became more muddled: the volunteers forgot 65 percent of the words. The researchers also measured skin conductance responses during the free association task. You might imagine that the thought of being late for an important appointment while being stuck in traffic would be more stressful than thinking about being taken care of, but that's not what the researchers found. Conflicted sentences made the volunteers sweat more than the negative sentences did. The volunteers also took longer to answer when reacting to conflicted sentences. Something was going on in the brain that slowed things down. The researchers speculated that it was the process of repression that delayed the response. In Jungian terms, the conflict was being pushed down into the unconscious where it became part of the shadow.

THE EVIL TWIN

The shadow is found in what Jung called the *personal unconscious*, a part of the unconscious that's closer to consciousness than the archetypes, which are universally shared. The reason the shadow is closer to consciousness is because it contains things that were once conscious and then became repressed. Old memories we think we have forgotten can also be found in the personal unconscious. Like the shadow, these memories of childhood can burst into consciousness when we least expect it.

The shadow is the first thing we encounter when we begin to explore our unconscious mind. It's the gatekeeper of the deeper regions where the magic of the archetypes resides. To access those archetypal regions, we've got to get past the shadow. Or to be more precise, we have to integrate the shadow into our consciousness: accept it as a part of who we are, join forces with it, and gain control over it. Like the archetypes, the shadow can possess us and make us behave in uncharacteristic ways, ways that may be abhorrent to us.

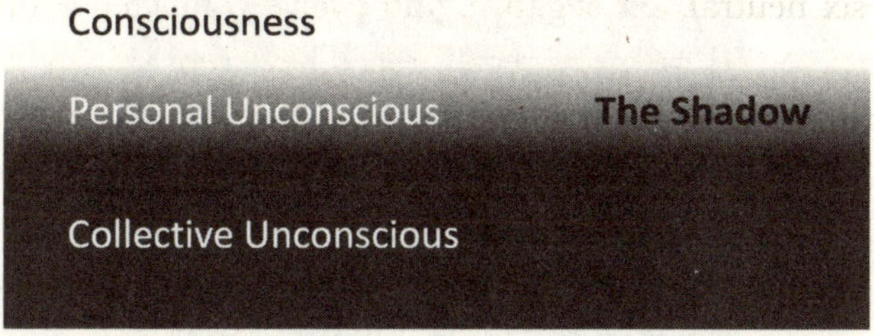

The shadow is found in the personal unconscious, which contains things that were once conscious, in contrast to archetypes, which are part of humanity's collective heritage.

But like mythological spirits and demons, the shadow can also give us special powers, powers of strength, endurance, and energy. Being able to work with the shadow is essential to growth.

The idea of an evil shadow lurking within seemingly ordinary people is a popular mythological motif. One example is Robert Louis Stevenson's story *The Strange Case of Dr. Jekyll and Mr. Hyde*. Dr. Henry Jekyll is a respected doctor, known for his decency and charitable works, but he also has a frivolous, pleasure-seeking side, which he conceals. It makes him feel dishonest to hide this part of himself, so he concocts a drug to free himself of this burden by splitting his psyche into two parts, one all good, the other all bad. The bad part becomes the cruel, murderous alter ego, Edward Hyde.

In contemporary culture, the evil twin motif has become so common, for example in soap operas and in the Bizarro world of the Superman comics, that it's evolved into a kind of joke. But its popularity reflects its deep resonance within the human psyche.

One of the best ways to understand the shadow is to look at it in yourself. It's not difficult to find evidence of it. Let's take sadism. Ask yourself to what degree you take pleasure in witnessing the suffering of others. You might think the answer is "none"—you take no pleasure at all in witnessing the suffering of others. But if you're like most people, that simply isn't true. Sadism is an ordinary pleasure, which we indulge in regularly.

For instance, action movies predictably end with the villain's defeat. But it's not enough to just kill him or put him in jail. For the movie to be satisfying, the villain must suffer. His super weapon is blown up, and his impregnable

The shadow may be portrayed as an evil twin. (Top: Poster from the 1931 Paramount film *Dr. Jekyll and Mr. Hyde.* Bottom: Mr. Spock from the TV series *Star Trek.*) *(Source: Left:* Toronto Star*; right: Memory-alpha.fandom.com.)*

lair burned to the ground. It's a pleasure to watch his lifetime of work, his hopes and dreams, crushed before his eyes, as he dies a gruesome death.

A description of Marvel's *Iron Fist* is typical of the genre: "His sworn enemies control his business and his city. They think he's dead. They're about to be very disappointed."[2] His enemies may be disappointed, but the audience is going to be thrilled as the hero metes out their just deserts—a good beating at the very least.

Some of the most popular children's cartoons, such as *Tom and Jerry*, are almost nothing but sadistic fantasies. The purest example is the Road Runner cartoons. They have no story, just an endlessly repeated ritual of the sadistic torture of Wile E. Coyote. And despite the cartoons' repetitive nature, even adults find them delightful.

Taking pleasure in the suffering of others isn't confined to fictional fantasies. When it happens in real life, we call it *schadenfreude*, a German word that means "harm-joy." It's most often experienced when someone seen as immoral receives punishment, because it's important that the victim of our sadistic enjoyment deserve the suffering being inflicted on him. That allows us to relax the normal prohibitions against sadism our conscious minds erect. We get to enjoy the best of both worlds. Our sadistic desires are gratified while we maintain the conviction that we're decent people who are on the side of justice.

Those prohibitions are important because sadism can easily get out of hand. The popularity of the Roman circus, in which audiences cheered the spectacle of wild animals tearing criminals to pieces, is one example. The ingenuity of the human mind when it comes to designing instruments of torture is almost unlimited. Enjoying action films may be a gentle breeze compared to the wild tempest of taking pleasure in blood sports, but they blow from the same point on the compass. The magical world of the unconscious is a dark place, and the pitfalls are abundant. As Jung writes, "Anyone who has to do with the phenomena of the unconscious knows with what hair-raising irrationalism and with what shocking tactlessness and ruthlessness the unconscious 'mind' dismisses our logical concepts and moral values."[3] We must tread carefully.

We achieve wholeness not by splitting off unconscious impulses, as Dr. Jekyll did, but by integrating them into consciousness where we can control them. We must sail our ship between the Scylla of denying our dark side and the Charybdis of giving in to it. Acceptance that it's an authentic part of who we are is the first step, because when something that is unconscious moves into consciousness, it's transformed. What we find in the unconscious is often wild and brutish; by making it conscious, we make it human. We subdue its unruly aspects, so we can use its power, the way a horse trainer turns a wild animal into a companion and helper. Unconscious sadism can become conscious fortitude, the strength of mind that enables someone to inflict harm when necessary.

Imagine a surgeon slicing through the skin to begin the process of opening a patient's chest. Every medical school has stories about students who have fainted watching this spectacle for the first time. But whether by putting in an intravenous line or finding the precise location of an injury during a physical exam, doctors must inflict pain, and it's best if they can do it without experiencing emotional distress. Most of us have met callous doctors, ones who've gone too far down that road, but which surgeon would you rather have: one who flinches in sympathetic pain as she makes the initial incision, or one who feels her excitement rising as she draws her scalpel through the flesh because she loves performing surgery?

People in any trade or profession can give examples of the need to inflict pain in the service of helping someone. Most of us have had to deliver bad news. And while there's no way to avoid giving pain, there are some ways that are better than others. Finding the best way and doing it well requires fortitude.

Some people are so uncomfortable with their dark side, it can almost paralyze them. The *Wall Street Journal* once published a story about one of the trials of modern life: paying for a cup of coffee using an electronic tablet that asks you for a tip.[4] Few people feel it's necessary to tip the person behind the counter who hands them a cup of coffee and a muffin, but some people have a hard time saying no:

> "It's so awkward," says [one of the customers] who taps 20%. "You press the middle button so you don't look cheap to the people behind you in line."
>
> When he complained to the company that makes the credit card reader, they said, "Tap 'No Tip.'"
>
> He responded, "I don't think I have it in me."

PROJECTION

One consequence of refusing to accept one's shadow is that it gets projected onto others. It's not unusual for individuals who are constantly angry to view themselves as reasonable people. Their anger is always the fault of someone else. As the familiar proverb tells us, if one person you know is a jerk, that person is probably a jerk. But if everyone is a jerk, the jerk is probably you.

Projecting an unwanted trait onto other people makes it easier to deny having it yourself. It's as if the spiritual contamination must go somewhere; it refuses to simply disappear. If it's cast out from one's inner world, it takes up residence in the outer world. Jung writes that "the effect of projection is to isolate the subject from his environment, since instead of a real relation to it there is now only an illusory one."[5] When people project, reality is distorted. What they believe is the outside world is actually a reflection of their own inner world.

To test the theory that projecting an undesirable trait makes it easier to repress, researchers recruited fifty-five college students into a study and gave them feedback about how dishonest they were.[6] The process began with a personality questionnaire, followed by a summary report given to the students in a sealed envelope. The students didn't know it, but the scores weren't real. Everybody was randomly given either 42 or 92 out of 100. At the bottom of the report was a narrative description of the result. The high scorers' description read:

> This person scores above the normal range of dishonesty or repressed guilt, indicating that this person may be at risk for dishonest behavior, especially when working under conditions that lack structure or supervision.

Next the students were told they would participate in an unrelated experiment that looked at how people used information to evaluate strangers. They read a vignette about a college freshman who purchased a music CD, copied it, then returned it to get his money back, claiming it was damaged. The students had to rate the freshman on fourteen traits, such as "boring," "intelligent," "likable," "lazy," and so forth.

Buried within the list were three traits that described a dishonest person: "deceitful," "a thief," and "a liar." However, only half the students were asked to rate these dishonesty traits. The other students received a list from which the dishonesty traits had been removed. Consequently, this group of students had no opportunity to make judgments about the dishonesty of the CD-copying freshman. (That will turn out to be important.)

The last thing the students in the study did was to fill out a questionnaire to rate themselves on various traits. One of those traits was dishonesty. Let's look at the results and see what happened.

The first question the researchers examined was whether the fake dishonesty report would affect the way students judged the CD copier. It did. Those who were told they had scored a 92 on the dishonesty scale were more likely to condemn the CD copier than those who were given a score of 42. Everyone has some degree of dishonesty; it's part of being human. Although they used deception to do so, the researchers forced the high-scoring volunteers to confront something unpleasant about themselves that lay just beneath the surface, something too difficult for many of them to face. As a result, they engaged in projection.

The next question was whether condemning the CD copier would have any effect on how the students rated themselves regarding dishonesty. If projection makes it easier to deny our own unacceptable traits, being given the opportunity to condemn the trait in someone else will make us less aware of it in ourselves. Sure enough, those students who were given the opportunity to rate the CD copier on dishonesty traits rated themselves as less dishonest. Those who were denied the opportunity to project, because the dishonesty traits had been removed from their list, gave themselves higher dishonesty ratings. In the absence of projection, people are more open to seeing the dark side of themselves, and the repressed contents within the shadow can be recognized.

The opportunity to gain control of the shadow isn't the only benefit of making it conscious. The amount of consciousness we have is not a fixed quantity. Newborn babies have a limited ability to maintain consciousness, spending most of the day sleeping. In adults, the capacity for conscious awareness varies from person to person. As we saw in an earlier chapter, some people are awake and alert for large parts of the day, while others go through life in a dreamy state, lost in thought and largely unaware of their surroundings. But we all experience fluctuations in consciousness to some degree; if you're a night owl, you might be more alert in the evening and less so when you first awaken, and vice versa if you are a lark. Because we do our best work when we're most conscious (and also tend to feel happiest during these periods), the more conscious we become, the more productive and fulfilled we are likely to be. Being fully conscious requires mental energy, and the unconscious—the ouroboros—is the source of mental energy, as it is the source of all nourishment and fertility. Incorporating pieces of the unconscious gives us access to their primal energy and increases our capacity to be awake and alert.

Shadow contents will express themselves one way or another. If we allow them to be consciously experienced, they won't need to paint those around us with their dark colors. The shadow contains the worst of us, and when we project it onto other people, it makes them seem hostile and frightening. When we own the projection—when we acknowledge it as part of ourselves—it returns to its origin and everyone appears to be friendlier and more open. People are more approachable, and we become more optimistic that our interactions with them will lead to good things. Owning a shadow projection wakes us up and makes us see more clearly. Encountering it for the first time may be mortifying, but if we are brave and don't avert our eyes, we'll be the richer for it.

BALANCING OPPOSITES

In the 1970s there was a popular TV show called *Kung Fu* that starred David Carradine as Kwai Chang Caine, a young Shaolin master and expert in advanced fighting techniques. Under ordinary circumstances, Caine was unassuming and soft-spoken, spending his spare time meditating or playing his bamboo flute. Many of the episodes in the series followed a similar pattern. Caine is tormented by ruffians and suffers their abuse humbly. Inevitably, the bad guys go too far, by threatening someone who is defenseless, and Caine thrashes them using his kung fu moves.

The show draws a connection between Caine's confidence in his fighting skills and his willingness to humbly accept being pushed around. Think of the *Star Wars* character Obi-Wan Kenobi, who's happy to be deferential to a drunken troublemaker in a bar because he knows he can cut the brute to pieces at a moment's notice with the lightsaber hidden in his cloak. Caine's hidden martial arts skills give him the self-assurance that allows him to be humble and gentle toward the people he meets, even those who provoke him. A good word to describe him is *meek*.

The word *meek* has an interesting history. Its most famous use is in the King James Version of Matthew 5:5: "Blessed are the meek: for they shall inherit the earth." The New Testament was written in Greek, and the word translated as *meek* is the adjective *praus*. We think of meek people as passive and docile, but *praus* points to a more complex meaning. It represents a balance between exercising power and showing restraint. In ancient Greece, the word was most commonly used to describe a horse trained for battle. A good warhorse must be both fierce and restrained, neither wild nor timid. It must fight savagely even while obeying every command of its rider.

The ancient Greek philosopher Aristotle classified *praus* as a virtue in his *Nicomachean Ethics*. Aristotle defined virtues as the mean between the extremes, the middle position between having too little (deficiency) of some quality and having too much (excess), which were both considered vices. Aristotle writes that the *praus* man took a middle position with regard to anger, in between the vices of timidity (a deficiency of anger) and combativeness (an excess of anger). The gentle, or *praus*, person is not someone who never gets angry, nor is he someone who gets angry over trivial matters. Instead, he possesses the ability to express anger at the right person, for the right reason, at the right time, in the right way. His passion isn't suppressed; that would lead to the

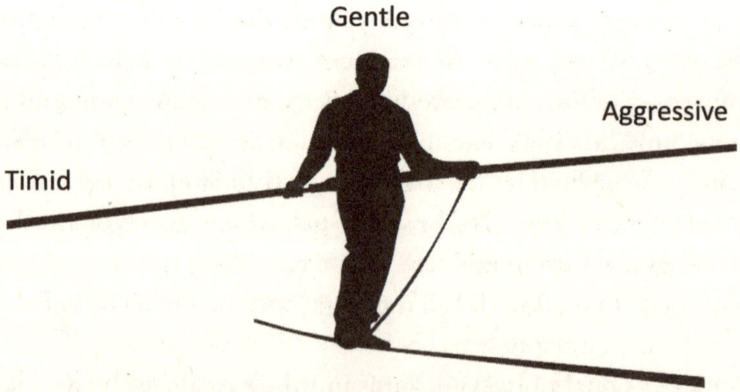

Integrating the shadow into consciousness allows us to use its strength to achieve balance.

deficient state of timidity. Rather, he places it under the control of rational judgment. Similarly, by making aspects of the shadow conscious, we achieve balance, finding the mean between the extremes of anemic rationality and wild instinct.

HARNESSING THE ENERGY OF THE SHADOW

We're all familiar with the mental energy the shadow can provide. Breaking the rules can cause anxiety, but often it's exhilarating. As the proverb tells us, "Stolen fruit is the sweetest." Violating an unreasonable prohibition makes us feel a little badass. Sneaking up on a roof to enjoy a forbidden view brings clear, focused attention as we enjoy the thrills of the present moment.

Consciousness must always fight to maintain its separation from the unconscious. Being fully alert in the present moment, as we've discussed, takes energy. But when we're feeling badass, it's effortless. The shadow drive that's been accepted by the ego and made conscious is accompanied by mental and physical energy.

The shadow may be most willing to help when we're breaking the rules, but it supplies its invigorating effects in other situations as well. A medical scientist at a large research university described to me how he once became infuriated reading a paper a colleague had written that directly contradicted the scientist's own research. In less than twenty-four hours, the scientist designed a study that would prove his colleague had made a serious error in his conclusions. Under ordinary conditions, designing a study of this nature took him weeks, but he was able to harness the power of his anger to dramatically speed up the process. (Doing so also helped him avoid angrily confronting his colleague, which in this case would have been counterproductive). It was almost as if his shadow did all the work for him, and all he had to do was sit back and watch it happen.

We see a fictional portrayal of this in the original Star Wars trilogy. It's widely known that director George Lucas was profoundly influenced by Campbell's *The Hero with a Thousand Faces*, and the trilogy contains many mythological motifs. The character of Darth Vader represents the shadow of the hero, Luke Skywalker. In the second film of the trilogy, Luke learns that he is inextricably bound to the man he thought was his enemy, because Vader is his father. This complicates the situation. Rather than just killing Vader, Luke needs to come to terms with him. The climax of this trilogy comes at the end of the third installment, when Vader is transformed from an enemy into an ally. At that point, Vader dispatches the real villain, the evil Emperor, while Luke watches from the sidelines. We realize that Luke's job wasn't to defeat the Emperor, but to befriend Vader.

Because this happens at the very end of this trilogy, for most of the story Luke is, in some ways, only half a person. Like Dr. Jekyll, he and his shadow (Vader) are two separate people, completely split off from each other. That makes Luke a little too good, a little bit flat, and certainly not the most interesting character in the series. That honor belongs to Han Solo, a more complex character who combines aspects of light and dark. Solo is portrayed as a scoundrel. He's a smuggler and sometimes a cold-blooded killer. In one scene, he blasts a bounty hunter who has confronted him. The scene is so dark that in a later release of *Star Wars*, director George Lucas recut it to make Solo fire in self-defense. (Solo fans were not happy.)

Solo accepts that he's a scoundrel, and you never know what he'll do: abandon you, show up when you need him most, or get you killed by flying into an asteroid field. Uniting the conscious and the unconscious, the process of individuation makes people more and more unlike everyone else, which makes them unpredictable—less like muffins, as Madeleine L'Engle would say. Other fictional characters who demonstrate a clear shadow side include Katniss Everdeen, Jack Sparrow, and Natasha Romanoff, the Black Widow—all of them more interesting, more individual, and more unpredictable than Luke Skywalker.

Integration of the shadow into consciousness may help us become more whole, but there are limitations to this integration. Another modern-day story filled with mythological motifs, the Harry Potter series, illuminates the nature of this limitation.

The reader sees aspects of Harry Potter's shadow distributed among three characters: Severus Snape, Draco Malfoy, and Voldemort. By the end of the series, Harry and Malfoy have saved each other's lives, and Harry, using magic, has seen the world through Snape's eyes and ultimately honored him as the bravest man he ever knew. The aspects of the shadow that these two characters represented could be integrated. But not every aspect can be tamed: the darkness that was in Voldemort had to be rejected and destroyed.

It's not easy to tell whether something is evil or merely looks evil because it hasn't been tamed by consciousness. Is the witch really a witch, or is she Mother Nature in disguise? Is the evil beast really a beast, or is it an enchanted prince who needs transformation? Who would have guessed that Vader would deliver Luke from the power of the Emperor? Identifying the parts of the shadow that are potentially helpful is crucial, as is being able to distinguish between what looks bad because it needs transformation and what is irredeemably evil.

We need the wild energy of the shadow (the parts that can be transformed into helpers) to successfully overcome what's truly evil inside us. Dumbledore, the headmaster of Harry's school, repeatedly emphasized that Voldemort couldn't be defeated without the help of Slytherin, home of the dark wizards. If the Slytherin students and teachers had been evicted, cut off from the rest of the school the way Mr. Hyde was cut off from Dr. Jekyll, the dark wizards would have joined Voldemort. But by accepting them, many—though not all—became allies. It's hard to know which parts of the shadow should be accepted and which ones rejected, and it's guaranteed we'll make mistakes before we get it right. Finding our true selves requires not only courage and strength but also a willingness to take risks. If you're too afraid of cuts and bruises, you will never learn to ride a bike. If you're too afraid of the demons inside you, you may never be able to claim your birthright of inner strength.

There's disagreement over whether humans are fundamentally good, though imperfect, or fundamentally evil, though capable of good acts. Perhaps the best answer is that we are neither. We can choose whether we will be primarily good or bad. But if we choose goodness, as most people do, it would be a mistake to blind ourselves to the evil inside us. If we deny its existence, we become powerless against it. Accepting some of that darkness can help make us strong, so that we can overcome the parts that cannot be accepted under any circumstances. We make Darth Vader an ally, so the Emperor can be destroyed.

The shadow may be frightening, but it's also understandable. It speaks the same language as the conscious mind because it's made up of rejected urges, attitudes, and feelings that were once conscious. Beyond the shadow is a deeper level of the unconscious, one that contains even more treasure and more danger. Unlike the shadow, it speaks a different language: the language of magic. To fully understand the human condition and discover the path that leads to our true selves, we first need to gain some familiarity with magic and the magical perspective on life. It is that task we turn to next.

Part II

MAGIC

FAIRY TALES

Fairy tales are the purest and simplest expression of collective unconscious psychic processes.

MARIE-LOUISE VON FRANZ

EINSTEIN'S ADVICE

Once upon a time the famous physicist Albert Einstein was confronted by an overly concerned woman who sought advice on how to raise her small son to become a successful scientist. In particular, she wanted to know what kinds of books she should read to her son.

"Fairy tales," Einstein responded without hesitation.

"Fine, but what else should I read to him after that?" the mother asked.

"More fairy tales," Einstein stated.

"And after that?"

"Even more fairy tales," replied the great scientist, and he waved his pipe like a wizard pronouncing a happy end to a long adventure.[1]

No one knows whether this story is based on an actual conversation or if it's pure fiction. It was first reported in 1958 in the *New Mexico Library Bulletin*, and over time, like any good fairy tale, it traveled and evolved into many different forms. One version that's found in collections of quotations attributed to Einstein is, "If you want your children to be intelligent, read them fairy tales. If you want them to be more intelligent, read them more fairy tales." The story became so popular that the Library of Congress researched its origins, but the results were inconclusive.[2] To this day, whether the conversation really took place remains unknown.

If the story is true, one wonders why the world's most famous scientist would have such a high regard for fairy tales. On the other hand, if the story is not true, one wonders why the idea that Einstein admired fairy tales should be so compelling that it spawned an urban legend. Later in this chapter, we'll look at some clinical research that appears to justify Einstein's legendary enthusiasm for these stories. But whether the advice came from the lips of the famous physicist or grew organically out of the soil of popular belief, it points to the importance of fairy tales in the development of the mind.

CROWDSOURCED LITERATURE

Although in modern times fairy tales are considered children's literature, they were traditionally called "household tales" and enjoyed by people of all ages. They were often recounted by women while they were spinning, sewing, or mending, giving rise to the expression "to spin a yarn."

One reason fairy tales move us is that they weren't written by any single individual. Rather, they originated in the oral tradition and were passed along from one storyteller to another, with each teller embellishing or omitting various details. That process allowed the tales to evolve in a way similar to natural selection. As the story was told to different audiences, details, themes, and plot points were added or taken away, sometimes without the storyteller even realizing it. When these changes resonated with an audience, they were maintained as part of the story, like a genetic mutation that gives a biological organism a survival advantage. When a modification failed to move an audience, it was left out of subsequent narrations. Different versions of the stories competed with one another, and the ones that were most evocative spread the widest.

Some fairy tales are thousands of years old and have traveled halfway around the world; one version of Cinderella has been traced back to ninth-century China, and there is a Greek version that goes all the way back to the sixth century BCE.[3] Fairy tales resonate with audiences across cultures and time because they have been stripped down to the bare essentials of what makes a good story. In them, the archetypes—the blueprints of the stories we use to make sense of the world—shine through in their clearest, purest form. Over time and distance, the stories become stronger, more effective representations of our common psychological foundation, the collective unconscious.

The collective unconscious is deeper than the personal unconscious, further away, and more alien to the conscious mind. The effects it can have on subjective experience and behavior—its magic—are more powerful and more dangerous because the collective unconscious can so easily overwhelm the ego. As our evolutionary inheritance, the collective unconscious is the home of the psychological forces of nature, the archetypes encoded within our DNA. When we come into contact with them, we feel as if we're grappling with something powerful that we can neither see nor control. Like the circuits that stimulate oxytocin secretion in a new mother, the archetypes motivate us to carry out their designs, and they have a profound influence over the course of our lives. Compared to the ego, they see the world from a radically different perspective, but it's essential that we understand the archetypes and learn to work with them. Fairy tales, ancient stories that have resonated with audiences for thousands of years, can help.

Like all products of the unconscious, fairy tales have both a light side and a dark side. Modern versions, such as the Disney adaptations, tend to emphasize the light side. But more traditional versions include scenes that are as dark as those found in horror films. The creator of the blood-soaked horrorfest *The Texas Chain Saw Massacre* said he envisioned it as an updated version of Hansel and Gretel.[4] In the fairy tale "The Goose Girl," the punishment of a waiting maid who betrays her mistress is to be "put in a barrel which is studded inside with pointed nails, and two white horses should be harnessed to it, which will drag her along through one street after another, till she is dead."

Clearly, fairy tales aren't for everyone, and they're best enjoyed in small doses. It can be tempting to pick up an anthology and read one story after another, but it's better to read just one, and then let it work on you for a while. Fairy tales aren't like other stories because fairy tales speak directly to the unconscious. They stir things up, making unconscious contents more likely to erupt into consciousness. A fairy tale tends to inhabit one's thoughts for a long time after it's been read. You may find yourself thinking about it for days. It might even intrude into your thoughts in ways that are uncomfortable. Some people dislike fairy tales because they activate frightening feelings that are hard to manage. (If that happens to you, respect what your unconscious is telling you and put the book away. There are gentler ways to approach the unconscious that we'll explore in later chapters.)

As we look at specific examples of fairy tales, we'll take a Jungian perspective, which understands them as descriptions of the process of individuation in which the ego and the unconscious come together to form the self. This goal might be represented by the discovery of a treasure, a precious stone, or the water of life. Most commonly, it's represented by a royal marriage that leads to the state of happily ever after. These stories anthropomorphize unconscious forces, helping us better understand their nature so we can learn to work together with them as equal partners.

THE UNCONSCIOUS IS DISGUSTING

The most famous collection of fairy tales in the Western world is the book of folklore collected by Jacob and Wilhelm Grimm—the Brothers Grimm. The Grimms' versions of popular tales are darker than other well-known retellings, such as the French versions by Madame d'Aulnoy and Charles Perrault. These more refined adaptations dress up the tales in courtly manners and wink at the reader with ironic humor, signaling that we shouldn't take the stories too seriously. The Grimms' versions, by contrast, are closer to the raw, unadorned contents of the unconscious, and therefore better suited to our purposes. They contain not only scenes of happily ever after but also depictions of the violence and brutality contained within the instinctual archetypes.

Grimm's fairy tales were written in German, and the translation we'll use is Margaret Hunt's, published in 1884. We'll begin with the first story in the collection, "The Frog Prince." It's a story about the difficulties two very different characters experience as they negotiate their relationship and the magical transformation that occurs when they ultimately unite. The story helps us better understand the task of individuation, in which the ego and the unconscious learn to accommodate each other and merge into the unity of the self.

The story begins:

> In old times when wishing still helped one, there lived a king whose daughters were all beautiful, but the youngest was so beautiful that the sun itself, which has seen so much, was astonished whenever it shone on her face.

The hero or heroine of fairy tales is often the youngest child. There are various ways to understand why this is. The first is as encouragement for young children. Growing up is difficult, and small children may feel powerless compared to adults and older siblings. Fairy tales in which the youngest succeeds where older, more powerful characters fail tell them they are important, and that they will prevail if they have courage and perseverance.

In many fairy tales, the hero is not only the youngest, but he's also the stupidest. His nickname may be "Dummling" or "Simpleton." That leads us to a second interpretation: a reminder that coming to terms with the unconscious is not an exercise of intelligence. Although analytical intelligence is a favorite tool of the ego, the simpletons of fairy tales rely on kindness and trust rather than cleverness. They share what little they have, and they listen to the advice magical creatures give them.

The youngest child can also symbolize the ego. From an evolutionary perspective, the ego is the part of the mind that developed most recently. Compared to the ancient unconscious, it's young and weak. But it's not to be looked down on, because, despite its frailty, in this story it's symbolically characterized as beautiful. Consciousness is unlike any other product of natural selection. Consciousness opened our eyes, allowing us to see beyond mere survival and reproduction. It allowed us to transcend our biological roots, giving us the ability to contemplate higher things, such as justice, beauty, and truth.

The beauty of the princess is so great that the sun itself is astonished when it shines on her face. Recall that light is a symbol of reason and consciousness, and the sun is the most brilliant source of light. Right from the beginning, we are warned not to underestimate this newcomer.

The story continues:

> Close by the King's castle lay a great dark forest, and under an old lime-tree in the forest was a well, and when the day was very warm, the King's child went out into the forest and sat down by the side of a cool fountain, and when she was dull she took a golden ball, and threw it up on high and caught it, and this ball was her favourite plaything.

A castle, since it's a human-made structure, represents consciousness. Nearby is "a great dark forest," ancient, obscure, and created by nature rather than humans. A forest is full of wild beasts and unexpected dangers. Unlike in a city, with its well-marked roads, one can easily get lost in a forest. There is danger, even mortal danger, but forests are also full of life. Soil, water, and air are transformed into living creatures. It's a representation of the ouroboros.

The first thing we encounter in this forest is a fountain. Fountains transport water from inside the earth up into the light and symbolize a channel of communication between the ego and the unconscious. In this story, the barrier separating the ego and the unconscious is thin. The forest is close to the castle, and it contains a fountain. There are indications that, in this story, achieving individuation will be relatively straightforward.

The next thing to enter the story is a golden ball. The sphere is a symbol of the self, because it's a shape that represents unity and perfection. While playing with the ball, the princess accidently drops it down the well and begins to cry. It's odd that the symbol for the self would be lost in the depths of the unconscious. The self is supposed to unite and therefore contain both the ego and the unconscious. This incident provides a clue that the golden ball isn't the princess's true self. It's a childish version she plays with as a toy, the way children play with dolls to prepare themselves for parenthood. The princess is in the process of growing up. As a young child, finding her self (the golden ball) was easy, but eventually it becomes submerged in the unconscious. The princess will have to work and grow to win it back.

A frog hears the princess crying and offers to retrieve her golden ball. Why is her helper a frog? It's certainly not the noblest animal. In fact, frogs are slimy and disgusting. But that's the point. At first, we're repelled by the contents of the unconscious—the smelly, slimy, ferment of our animal nature. The frog is introduced as "stretching forth its thick, ugly head from the water," and its ugliness is emphasized throughout the narrative. The ancient unconscious may regard the beauty of youthful consciousness with awe, but the admiration isn't mutual. Attaining wholeness and becoming the person we were meant to be is our heart's desire, but at the beginning, what we find in our unconscious appears ugly. Our first reaction is revulsion.

Another reason her helper is portrayed as a frog is that frogs are amphibious; they live both on land and underwater, spanning two worlds. That makes the frog a good symbol for the transcendent function, the drive toward completeness in which the realms of the conscious and unconscious unite. It's neither a fish nor a land animal. It's both, and it will not only retrieve the princess's toy self, it will also lead her to her true one.

The frog asks her, "What wilt thou give me if I bring thy plaything up again?" One cannot attain wholeness without making sacrifices. The princess answers that she will give it anything and everything.

> "Whatever thou wilt have, dear frog," said she—"my clothes, my pearls and jewels, and even the golden crown which I am wearing."

The princess is not yet fully mature, but this answer reveals her inborn wisdom. To attain the most valuable thing in the world, her true self, she understands she must be ready to sacrifice everything, like the merchant in the Gospel of Matthew who sold all he had to buy the pearl of great price. However, although the princess is on the right track, she doesn't fully understand the sacrifice she will need to make. The frog rejects the things she offers, telling her,

> "I do not care for thy clothes, thy pearls and jewels, or thy golden crown, but if thou wilt love me and let me be thy companion and

> play-fellow, and sit by thee at thy little table, and eat off thy little golden plate, and drink out of thy little cup, and sleep in thy little bed—if thou wilt promise me this I will go down below, and bring thee thy golden ball up again."

The price of wholeness is accepting the disgusting frog as her companion. The princess promises, and the frog dives into the well and returns with the golden ball.

Before we move on with the narrative, it's worthwhile to look at the frog's demands more closely because they contain some strange things. Despite saying it wants to be her companion, the frog asks for things that are quite a bit more intimate than simple companionship would suggest. Friends don't typically sleep together in the same bed. That's more characteristic of lovers. Friends don't eat off the same plate or drink from the same cup. Not even lovers do that. The frog is indicating that the relationship it has in mind goes well beyond friendship, beyond even romantic attachment. By describing the two of them eating off the same plate and drinking from the same cup, the frog implies they will unite to become a single individual.

The princess agrees to the frog's terms, but as soon as she gets her ball back, she loses interest in the frog and runs away, with no intention of keeping her promise. By breaking her promise, she demonstrates a lack of integrity, and that's a problem, because the ego must be strong before it can join with the unconscious. Most fairy tales require the hero to undergo a series of trials to demonstrate a strong, stable ego before achieving the unified state of "happily ever after."

The next day, while the royal family is having dinner, there's a knock at the door.

> The princess ran to see who was outside, but when she opened the door, there sat the frog in front of it. Then she slammed the door to, in great haste, sat down to dinner again, and was quite frightened. The King saw plainly that her heart was beating violently, and said, "My child, what art thou so afraid of? Is there perchance a giant outside who wants to carry thee away?" "Ah, no," replied she, "it is no giant, but a disgusting frog."

The transcendent function may appear small and contemptible, but it's as frightening as a giant because we know it has the potential to upend our lives and transform us.

At this point in the story, the princess is facing difficulties, but let's not underestimate the trials of the poor frog. Getting to the dining hall required some perseverance. The small animal had to hop all the way from the woods to the palace and then climb up the palace stairs. Even after the princess slams the door in its face, it stubbornly remains where it is, refusing to leave. The transcendent function is persistent. If the ego doesn't cooperate willingly, the transcendent function will chase it down.

The king asks his daughter what the frog wants, and she confesses the promise she made. The king tells her, "That which thou hast promised, must thou perform. Go and let him in." Keeping a promise is an important strength of a rational ego, which processes information based on universal rules. An unconscious, experiential approach, by contrast, views things individually, which makes each situation unique. From that perspective, the princess's relationship to the frog in the castle is different from what it was by the well. But the princess must keep her promise despite the change in the location and in her feelings. She must exercise the ability of the ego to work with abstractions that hold true regardless of the extraneous details that surround them.

The princess lets the frog in. It hops over to her chair and says, "Lift me up beside thee." The princess must raise up the repellent unconscious and make it her equal. Encouraged by her father, she does so, allowing it to eat with her, and when they have both finished their dinner, she takes it upstairs to her bedroom. She places the frog in a corner, but it tells her it wants to get in bed with her.

What happens next in the Brothers Grimm version is different from other, more popular versions of the story. In the more familiar version of the story, the princess transforms the frog by kissing him. That's a perfectly good way to move the story forward, showing that the princess learns to overcome her aversion to her unconscious. But that's not where the Grimms' version takes us. After the frog demands to be put in the princess's bed, the tale continues, "Then she was terribly angry, and took him up and threw him with all her might against the wall."

Why was this act of aggression necessary? Earlier in the story, not only had the king told the princess that she must open the door to the frog, he had also commanded her to lift up the frog when it wanted to sit next to her, and bring it upstairs to her room when it said it was tired. He acted as a sort of auxiliary ego, explaining to the princess one of the rational rules of moral behavior: "You should not despise someone who has helped you in time of need." Keeping the promise she made was only accomplished with outside support. Children need that kind of support, but ultimately the ego needs to learn to act on its own initiative, and that may first require rebellion against the authority that seeks to control it. The ego may need to draw on the power of the shadow to accomplish that rebellion, as the princess did when she hurled the frog against the wall.

Whatever the reason, the raw aggression she displays is transformational. "When he fell down he was no frog but a king's son with beautiful, kind eyes." The redeemed prince tells the princess that he had been turned into a frog by a witch, and the princess was the only one who had the ability to deliver him from his curse. Only your ego can raise up your unconscious into consciousness. Although the archetypal blueprints are part of the collective unconscious shared by all humanity, each individual expression of the blueprint is unique, and each one requires a unique ego to make it whole.

MIRACULOUS HELP

With its enormous processing power and instinctual energies, the unconscious can be a powerful ally. We've all been surprised by strengths and abilities we didn't know we had. We've looked back on past accomplishments and wondered how we ever managed to pull them off. When the unconscious joins forces with the ego, we're capable of performing incredible feats of mental strength, endurance, and creativity. There are few things in life as valuable as gaining the cooperation of the unconscious, and stories about magical creatures who give the hero or heroine the abilities they need to overcome impossible odds are common in collections of fairy tales.

One well-known story of magical help is "The Shoemaker and the Elves," which begins with a poor shoemaker on the verge of starvation: "A shoemaker, by no fault of his own, had become so poor that at last he had nothing left but leather for one pair of shoes."

Magical help generally comes to those in need. The rich and comfortable view themselves as self-sufficient, and they're rarely visited by magical creatures. The ego resists the unconscious and avoids contact with it as long as it can. It is only when we get into a bad situation, one that we can't get out of on our own, that we become desperate enough to open ourselves to the influence of our unconscious. Necessity is not only the mother of invention, but it's also the mother of growth.

The shoemaker takes the last of his leather, cuts it into pieces to be assembled into shoes, and goes to bed, planning on sewing the pieces together in the morning. When he wakes up, the shoes are not only finished, but "They were so neatly made that there was not one bad stitch in them, just as if they were intended as a masterpiece."

Sometimes agents within the unconscious will do our work for us, and when they do, they generally do that work well—far better than the ego could. In the Brothers Grimm's "The Three Feathers," an ugly, subterranean toad produces "a ring that glistened with precious stones and was so beautiful that no goldsmith on earth could have made it." In Germanic folklore, dwarfs are portrayed as short, ugly creatures who live underground. But despite their uncouth appearance, they are expert craftsmen who produce exquisite artifacts of magical power. Agents within the unconscious appear crude and uncivilized to the ego, but they are capable of works of subtlety and beauty.

The shoes in "The Shoemaker and the Elves" are so well made that the customer who buys them pays more than the usual price, and the shoemaker is able to buy twice as much leather for the next day's work. Once again, he wakes up to find expertly made shoes waiting for him on his workbench. This goes on for some time, until one night the shoemaker and his wife decide to stay up to find out who's helping them. They hide in a corner of the room and see two little naked men sit down at the workbench and work all night until everything is finished.

The little men are naked for the same reason that animals are naked. Clothes are connected to consciousness, and the little men are symbols of the unconscious. We see the connection between clothes and consciousness in the book of Genesis. When Adam and Eve ate the fruit from the tree of the knowledge of good and evil, their eyes were opened, and the first thing they did was make themselves clothes. They became conscious of the fact that they were naked.

The shoemaker and his wife feel sympathy for the little men. They think they must be cold without clothes. The wife says, "The little men have made us rich, and we really must show that we are grateful for it." So they make them each a suit of clothes. Gratitude and sacrifice are the usual way wise characters in fairy tales respond to helpful or needy creatures. Often it involves sharing food with a beggar who is a powerful wizard in disguise, or it may take the form of rescuing an animal in distress. Sometimes a contrast is set up in which the brothers or sisters of the hero are hard-hearted, whereas the hero is kind and generous.

When the little men find the clothes, they are overjoyed. They put them on, dance around the workshop, and are never seen again. Losing the help of the little men may seem like poor compensation for the kindness of the shoemaker and his wife, but in the last sentence of the story we're told, "As long as the shoemaker lived all went well with him, and all his undertakings prospered." By clothing the little men, he made the unconscious conscious. No longer was he a passive recipient of supernatural help. The abilities that had been hidden were brought into the light, dressed in the clothes of consciousness, and became his own. Now the shoemaker possessed the skills of the little men, and all his undertakings prospered.

OUR ANIMAL NATURE

It's nice to get help sewing shoes when you're on the brink of starvation, but magical help is capable of much more. Fairy tale characters are often given impossible tasks: moving a mountain or retrieving a pearl from the bottom of the sea. Those who try and fail are punished with imprisonment or death. Pursuing our true self is far from safe. However, if we choose security, the result is stagnation. Another frequent consequence for failure is being turned to stone. The alternative to growth is petrification.

In pursuing the overwhelmingly difficult tasks they are given, characters who are kind and generous always receive some form of magical help.

When our backs are up against the wall, when the challenges we face seem impossible, we often find a strength we didn't know we had—a strength that gives us the ability to successfully fight our way through the problem. These magical helpers often take the form of animals, generally talking animals, which makes them part human, part beast, like the instinctual nature of the unconscious. They perform tasks that would be impossible for a human. In the Brothers Grimm version of "Cinderella," the heroine's wicked stepmother pours a bowl of lentils into the ashes and tells Cinderella she cannot go to the ball unless she picks them all out within two hours. Pigeons and turtledoves come to her aid.

The theme of magical helpers gathering together something that has been scattered is seen in other fairy tales, too. In "The Queen Bee,"

Birds come to the aid of Cinderella when her stepmother gives her an impossible task. The artist captures the disordered efficiency of the wild birds. (Alexander Zick, *Aschenputtel*, 1896.) *(Source: Wikimedia Commons.)*

thousands of pearls that have been hidden in a forest must be recovered in order to disenchant a castle. Two proud princes try, fail, and are turned to stone. The youngest, stupidest prince doesn't even try. "It was so hard to find the pearls, and the job was so tiresome!—so he sat down upon a stone and cried." But earlier in the story he had saved a colony of ants, and the ants find every pearl and stack them up for the prince in a neat pile. The human psyche may be full of fragmentary personalities, but if the ego can establish good relations with its hidden partners, the pieces can be gathered together to form a unified whole.

Sometimes the solution a magical animal offers isn't accepted by the hero. It can take time to establish trusting relationships among the different parts of the psyche. This is a problem that bedevils the characters in "The Golden Bird." In this story, three princes go off in search of a bird with miraculous golden feathers. Along the way, all three brothers meet a talking fox who begs them not to shoot it, offering good advice in exchange. The two older brothers reject the offer and try to kill it, but the youngest one, who is good natured, says, "Be easy little fox, I will do you no harm."

The fox tells the boy to sit on its tail, which allows the boy to travel as fast as the wind. The fox and the boy race along until they arrive at a small village where there are two inns on opposite sides of the road. One of them is brightly lit, full of people having a good time. The other one is shabby and dark. The fox tells the boy to spend the night in the shabby one. Although the two older brothers had tried to kill the fox, it had, nevertheless, given them the same advice. But they thought it was nonsense.

> "I should be a fool indeed," [the eldest son] thought, "if I were to go into the shabby tavern, and pass by the good one." So he went into the cheerful one, lived there in riot and revel, and forgot the bird and his father, and all good counsels.

The clever older brothers scorned the fox's advice: "How can a silly beast give wise advice?" But the simple, youngest brother did what the fox recommended and didn't get stuck, as his older brothers did.

The next morning, the fox tells the boy about a castle where he will find the golden bird inside a wooden cage. Next to it, the fox says, will be a golden cage, but under no circumstances should the bird be moved from the wooden cage to the golden one. Although the boy obeyed the fox the night before, in this instance, he disobeys. He reasonably concludes, "It would be absurd if I were to leave the beautiful bird in the common and ugly cage." The result of his disobedience is that the boy is captured and sentenced to death. But the king who owns the bird tells the boy that his life will be spared if he brings the king a golden horse that can run faster than the wind.

The young prince is dismayed by this seemingly impossible demand, but as he sets off in search of the golden horse, he meets the helpful fox again. "Look you," the fox says, "this has happened because you did not give heed to me. However, be of good courage. I will give you my help, and tell you how to get to the Golden Horse." The fox gives him instructions, warning him that when he finds the horse, he must saddle it with a plain saddle he will find there, rather than a golden one that will be next to it. The boy again disobeys him: "It will be a shame to such a beautiful beast, if I do not give him the good saddle which belongs to him by right." And again, he is captured and sentenced to death, but told his life will be spared if he can complete yet another task: he must bring back a beautiful princess living in a golden castle.

The boy's desire to put the golden saddle on the golden horse makes as much logical sense as did his desire to put the golden bird in the golden cage. The ego likes things to be ordered, with the right things in the right place. But the unconscious doesn't work that way. The fox is telling the boy that consciousness cannot stand on its own. Gold is associated with the sun, which is a symbol of consciousness. The golden saddle and the golden cage may appear most beautiful, but by pairing them with the golden animals, they create a situation that is too one-sided. The golden cage and saddle don't produce the necessary balance.

The fox is getting tired of the boy's refusal to follow his advice and tells him, "I ought only to leave you to your ill-luck." But once again he tells the boy what to do. Once again, the boy disobeys and is imprisoned under threat of execution. The final task he is given is to remove a mountain that's blocking the king's view. The boy takes a shovel and digs for seven days, but he can't make any progress on the mountain. This time, the fox doesn't bother giving advice; it simply does the job for him. The boy goes to sleep, and when he wakes up, the mountain is gone. This is a turning point. The boy begins to follow the fox's advice, and things go better. He becomes the owner of the bird and the horse, rescues his two brothers, and marries the princess.

The animal aspects of human nature have a kind of wisdom that can be difficult for the ego to accept. It's inevitable that there will be disagreement and conflict among the different parts of the psyche before harmony is established. But that's all right. We don't have to get it right the first time. It would be unrealistic to expect it. The fox became frustrated with the boy but didn't abandon him. Growth requires repetition. We get things wrong many times before we get them right. That's not to say that failure doesn't bring suffering. Each time the boy disobeyed the fox, he was imprisoned and sentenced to death. Failure can feel like we've hit a dead end and there's no way out. But if we don't give up, help will come, giving us more chances, until we finally succeed.

The last request the fox makes of the boy is an unexpected one. It occurs years after the boy has married the princess and become heir to the throne. One day, while walking in the woods, he encounters the fox, who begs him to shoot it dead and chop off its head and paws. Many years ago, the boy had refused this same request, feeling that it would be a cruel return for all the help the fox had given him. But this time he obeys, and the fox is transformed into the brother of the princess. He had been cursed by an evil spell and was finally free. Individuation begins with establishing friendly relations with the animal instinct. But it's not complete until the animal instinct is raised up and made human. The unconscious must become conscious. The fox had to be killed so it could be reborn as a man.

This is the second time we've seen the hero of a fairy tale act with violence toward a friendly animal. Originally, we interpreted the princess's aggression in hurling the frog against the wall in "The Frog Prince" as a way of asserting her independence. Now we see another way of looking at it. A certain amount of violence on the part of the ego is needed to humanize unconscious contents. They can't be integrated into consciousness in their natural form. They have to be reformed before they can be compatible with the ego. As we delve deeper into the magical tradition, we'll see an increasing emphasis on gentleness and the importance of compassion for the ancient, hidden parts of the self. But any kind of change involves the destruction of old ways of being, and so violence is an appropriate metaphor. The use of violence also speaks to the need for strength. Taming the wild instincts isn't easy. Highly developed ego strength is necessary. The metaphor of violence has its limits, however, as this type of violence doesn't establish an oppositional relationship. The humanized animals are always grateful, and the transformation the violence conjures leads to a closer relationship with the hero.

WHEN THE UNCONSCIOUS IS TOO STRONG

Our animal nature can help us accomplish tasks the ego can't manage on its own, but it can also overwhelm us. Our capacity to integrate its power into consciousness is limited. Too much can be our undoing, and being overwhelmed by power that one is not yet ready to wield is a common mythological motif. In the Disney cartoon *The Sorcerer's Apprentice*, Mickey Mouse tries to use a sorcerer's magic to fetch water, but he quickly loses control over the power he summons.

Before they can safely use magical power, characters must first train themselves. This training usually involves discipline—strengthening the ability of the ego to effectively manage its relationship with the unconscious, like a rider learning to manage a horse. In ordinary life, ego strength is reflected in the ability to tolerate frustration, regulate emotions, and resist impulses. Those who lack ego strength are at risk of being overwhelmed by their instincts and thus losing part of their humanity. We see such a transformation in the fairy tale "Little Brother and Little Sister," a story that contrasts the weak ego strength of Little Brother with that of his stronger sister. As the story begins, Little Brother and Little Sister are driven from their home by their cruel stepmother. Forced out of their familiar environment, they find themselves deep within the woods.

The next part of the story focuses on Little Brother's thirst. He wants to drink from a brook, but his stepmother is a witch, and she has cast an evil spell on all the brooks in the forest. The first one they find tells Little Sister that anyone who drinks from it will be turned into a tiger. She begs her brother not to drink, or else "you will become a wild beast, and tear me to pieces." He restrains himself, saying he will wait for the next opportunity. The second brook they come to tells Little Sister that anyone who drinks from it will become a wolf. When she expresses her fear that Little Brother will devour her, he once again restrains himself. But at the third brook, which has warned Little Sister that its waters turn people into deer, Little Brother can restrain himself no longer.

In some ways, this story is the opposite of "The Golden Bird." There, the hero had to learn to obey his instincts. Here, the boy must learn to resist them. In real life, we face a combination of the two problems. Sometimes, we overindulge the instincts that come from our unconscious, and the ego must learn to resist them in order to maintain its full humanity. At other times, we unwisely ignore our instincts, and the ego must learn to set aside its purely rational judgments and join forces with its ancient partners. We need to develop the flexibility to resist or accept instinctual impulses based on our circumstances. In "Little Brother and Little Sister," the waters are bewitched, and, therefore, thirst must be resisted. It's only after discovering what our task is (resist or obey) that we can apply ourselves to successfully accomplishing it, and that task will be different for each person in each situation.

Little Brother's task is to show restraint. He doesn't have as much ego strength as his sister, but by successfully resisting temptation twice, at least he avoids the fate of turning into a ferocious beast that would kill her. Instead, he becomes a harmless deer.

Little Sister cries when she sees her brother transformed into an animal, and she promises that she will never leave him. She undoes her belt and ties it around the deer's neck. Then, she gathers reeds, braids them into a rope, and fastens it to the belt. From one perspective, it doesn't look as if she's treating her brother very kindly. By fastening a collar and rope around his neck, she's depriving him of his freedom. But from another perspective, she is taking the first step toward restoring her brother's humanity. She is beginning the process of taming the wild animal. Just as the clothing the shoemaker and his wife sewed for the little men allowed them to become human enough to be integrated into the conscious ego, the halter made by Little Sister raises the deer above its purely natural state and into a closer relationship with consciousness.

Little Sister is the right person to help her brother. Her well-developed ego strength not only allows her to resist the enchanted waters, but also makes possible a harmonious relationship with her unconscious, symbolized by the natural elements of the forest. She is able to hear and understand the brooks when they warn of the dehumanizing effects of drinking their waters. This same strength allows her to tame the animal her brother has become, forcibly restraining it while treating it with love.

The theme of taming a wild animal also plays a prominent role in Antoine de Saint-Exupéry's *The Little Prince*. After traveling across the galaxy, the prince arrives on Earth, where he meets a fox, who asks the prince to tame it. The fox explains that in its untamed state it is just like every other fox in the world. It's undifferentiated, as Jung would say. But if it is tamed by a human, then it becomes an individual:

> "To me, you are still nothing more than a little boy who is just like a hundred thousand other little boys. And I have no need of you. And you, on your part, have no need of me. To you, I am nothing more than a fox like a hundred thousand other foxes. But if you tame me, then we shall need each other. To me, you will be unique in all the world. To you, I shall be unique in all the world."[5]

Taming not only makes a wild animal unique, but it also makes it *praus*, both fierce and obedient at the same time. The animal instinct retains its power, but now that it's been made conscious, the ego can work with it. Taming raises the wild instinct out of the darkness of the unconscious and makes it intelligible to the ego. The fox tells the Little Prince, "One only understands the things that one tames."

WHY READ FAIRY TALES?

Now that we can identify the theme of individuation that lies at the heart of many fairy tales, is there any point in taking the trouble to read them? Once again, we find ourselves in the territory of Walt Whitman's learn'd astronomer. It's one thing to look at the charts and the mathematical formulas. It's another thing to gaze at the stars. An intellectual interpretation of fairy tales is not the same thing as actually reading them, and the reason for this difference is that there is something special about the way stories affect the brain.

The importance of stories is first seen in childhood. Psychologists theorize that children learn moral behavior through narrative rather than by internalizing abstract rules.[6] Rules provide generalizations, which can be difficult for young children to understand and apply. Stories, by contrast, illustrate moral rules via characters the child can identify with. When faced with a specific moral problem, such as whether to tell the truth about breaking a dish, a child who loves the movie *Frozen* doesn't ask *What do the rules say?*, but rather *What would Anna do?* The child uses experiential instead of rational processing.

Identifying with a character is a complex process. It requires a child to empathetically experience the trials and tribulations the character undergoes in the story. It's both an intellectual exercise—understanding the meaning of the story—and an emotional one—sharing the internal reactions of the protagonist. In this respect, growth later in life is similar to the moral awakening of the child, in that intellectual understanding of the process of individuation is not enough. "Only emotional experience liberates," Jungian scholar Jolande Jacobi tells us. "It alone can bring about the necessary revolution and transformation of energies."[7]

Emotional experiences are different from intellectual understanding. As Jung writes, "The difference amounts roughly to that between a severe illness which one reads about in a textbook and the real illness which one has."[8] A story lies somewhere in between intellectual understanding and actual experience. When we hear a story, we feel vicariously the pain, the joy, and the other emotions of the characters. We are taken beyond cold facts and purely intellectual interpretations into the realm of the emotional experience Jung believed essential for growth.

How authentic is the vicarious experience that one gets from reading a story? From the perspective of neuroscience, the answer is *quite a bit.* When people hear a story, the brain of the listener mirrors the brain of the story's protagonist. For example, if you're listening to a story about someone running away from an attacker, the circuits in your brain that coordinate running are activated. If you hear a story about someone suffering pain, pain processing circuits are activated. That's why it makes you cringe when your friend tells you about the time he tripped and bent his knee backward. Vittorio Gallese, a professor of psychobiology at the University of Parma, writes that when we listen to a narrative, our brains generate "a physical, and not simply 'mental,' experience of the mind, motor intentions, emotions, sensations, and lived experiences of other people."[9]

This phenomenon has been described as a "mind meld" between storyteller and listener or, more formally, "neural entrainment." When researchers scan the brains of two people having a conversation, they find that when one of them is telling a story about an experience they had, the activity of the listener's brain mirrors what is going on in the storyteller's brain.[10] This neural entrainment doesn't occur with any other form of oral communication. It only happens with storytelling. Stories affect the human brain in a way that can't be duplicated by anything else. Which is why it's not enough to read *about* fairy tales; one must read the fairy tales themselves.

Stories also make us better mind readers, which helps us function better in groups. Mind reading is often portrayed as a supernatural phenomenon, but we do it all the time. We just call it "putting yourself in someone else's shoes," or more simply, "empathy." The best clue as to what another person is thinking is usually the expression on their face, and the area around the eyes is particularly informative. The eyes are so expressive that psychologists can even measure your ability to infer the mental state of others by showing you a photograph of just a person's eyes, then asking you to choose which emotion is being conveyed. It's called the Reading the Mind in the Eyes Test, or RMET for short.

A group of scientists gave the RMET to eighty-six volunteers after asking them to read a short text.[11] Half the volunteers read fiction, and the other half read nonfiction. The researchers found that those who read fiction before taking the RMET scored significantly higher than those who read nonfiction. Stories facilitate interpersonal interactions by making us better able to understand how other people feel. In other words, stories boost empathy.

A separate study found that the magnitude of this empathy boost was directly related to how engaged the readers became in the story.[12] Sometimes,

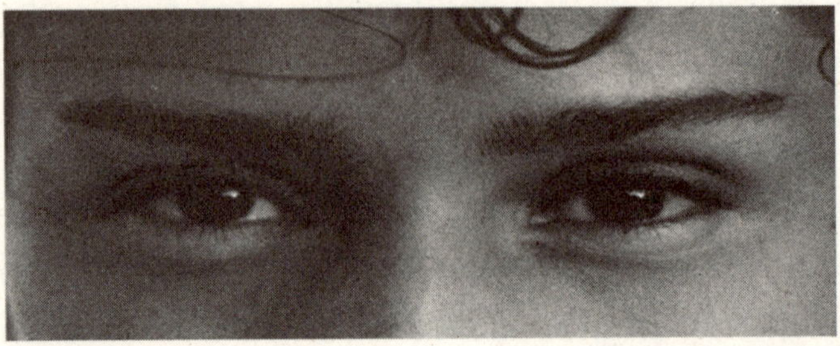

Playful | Comforting

Irritated | Bored

A sample question from the Reading the Mind in the Eyes Test. Can you determine the emotion of the individual in the photograph by looking only at his eyes?
(Source: Courtesy of the Autism Research Centre.)

when you read, all you see is words. At other times, you see pictures in your imagination that depict the events in the story. The researchers used this phenomenon to measure engagement with the story, rating each reader based on how vivid their mental imagery was. They found that the more engaged a reader was in a story, the more empathy he had for the characters. The effect carried over to real-world behavior. As part of the experiment, one of the scientists "accidently" dropped his pen, and then recorded whether the research volunteer picked it up for him. They found a significant correlation between engagement in the story and subsequent helping behavior.

We can gain insight into the mechanism by which stories promote altruistic behavior by looking at changes in the level of oxytocin, a hormone that focuses our attention on relationships, in a person's blood.[13] Emotional involvement with a story increases the hormone's secretion. In one study, an elevated level of oxytocin while watching a video about a child with terminal brain cancer was associated with greater concern for the people in the video. This elevated level of oxytocin was also connected to a greater willingness to help out in the real world, measured by how much of their own money the research volunteers agreed to donate to St. Jude Children's Hospital.

Reading stories takes us out of our own heads and puts us inside the heads of the characters in the narrative. We see things from another perspective, which inspires us to put the well-being of other people above our own, making us more effective members of our community. But reading stories also benefits us as individuals. Becoming entranced by a story— "lost" in a book—can lead to an altered state of consciousness, according to Brian Sturm, a professor in the School of Information and Library Science at the University of North Carolina at Chapel Hill. As part of a study designed to better understand the experience of becoming immersed in a narrative, Sturm interviewed participants at storytelling events.[14] Participants described "being transported" to a different place: "I'm no longer sitting in a tent listening to someone tell a story . . . I was in those woods, I saw those animals, they were real."

The altered state induced by stories has been compared to meditation and has some of the same benefits: less stress, greater self-esteem, and lower rates of depression.[15] According to a national survey of reading habits among adults in England, "People who read books regularly are on average more satisfied with life, happier, and more likely to feel that the things they do in life are worthwhile."[16] This isn't a recent discovery. The health benefits of reading have been known for thousands of years. The oldest-known library belonged to Pharaoh Ramses II, who reigned in the thirteenth century BCE. It had an inscription over its portals that designated it "the house of healing for the soul."[17]

Reading fiction might even be compared to the mystical experience. Sometimes the world created by an author can feel more real than ordinary reality. An expert storyteller calls upon mythological motifs to activate the unconscious, and these archetypal characters and events can become so compelling that, when the book is put aside, everyday reality seems pale in comparison.

Because of their magical content, fairy tales are more than just ordinary stories, and we might wonder whether they do more than simply increase our ability to connect with others. We've identified magic as the language of the unconscious. Is there any evidence that magical stories make it easier for the ego to also connect with its unconscious partner? Eugene Subbotsky, the same psychologist who demonstrated magical thinking in adults by "raising

Listening to stories can induce a trance-like state of consciousness. (George Dunlop Leslie's *Alice in Wonderland*, 1879.) *(Source: Wikimedia Commons.)*

the stakes," also undertook a study to evaluate one component of that connection: creativity, or the ability of the mind to develop novel solutions to problems that can't be solved using conscious reasoning. He found that when children were exposed to narratives that contained magical elements, it made them more creative than children who were given similar narratives without the magic.[18] Both groups of children experienced a story. But only one group experienced a magical story. And that turned out to be an important difference.

Subbotsky prepared the experiment by creating two fifteen-minute film clips using content from the movie *Harry Potter and the Philosopher's Stone* (renamed, as with the book, *Harry Potter and the Sorcerer's Stone* for the US market). He had a panel of judges who were unaware of the study's purpose ensure that both clips had the same emotional tone, pace, and richness in action. The only difference was whether the scenes contained magical content. Creativity was then measured using several standard creativity assessments, given before and after children watched the film clips. Before watching the clips, there was no difference in assessment scores between the two groups. After watching them, the average scores of the children who watched the magical version were significantly higher. On some measures their average score was more than double that of the nonmagical group.

Thinking creatively often involves solving problems by exploring many possible solutions, a practice called "divergent thinking." Divergent thinking is spontaneous, free flowing, and unorganized. It's typical of unconscious processes. By contrast, "convergent thinking" follows the logic of the ego to find one single "correct" solution. The ability to think divergently doesn't depend on intelligence as measured by IQ. Instead, it's found among people who are curious and persistent, who embrace nonconformity, and who are willing to take risks. Based on Subbotsky's work, we can now add to that list people who expose themselves to stories about magic.

Divergent thinking is a valuable skill to possess. Children who are adept at it are more likely to realize creative achievements as adults, such as awards, patented inventions, and published articles.[19] Perhaps this is our explanation for Einstein's purported enthusiasm for reading fairy tales to children as the best way to prepare them for careers as scientific pioneers.

Fairy tales are loved by people around the world. These crowdsourced stories strike deep. We enjoy them as children and remember them for the rest of our lives. They have such powerful effects because they evolved over the course of hundreds of years to become highly distilled, highly potent descriptions of the drama that accompanies the growing alliance between the ego and the unconscious.

Fairy tales are not, however, the only magical tradition sprung from our need to understand the fundamentally uncontrollable forces that steer our lives. Some of these magical arts were seen by their devotees not only as descriptive of those influences, but also as sources of power that could be harnessed to create wealth or discover hidden truths about the world. While we know today that those efforts were in vain, these traditions still have great value. The knowledge to be gained is not about the outer world, but about ourselves; the wealth to be unlocked resides within.

ALCHEMY

Projections change the world into the replica of one's own unknown face.

C. G. JUNG

CHEMISTRY AND MAGIC

Alchemy is an obscure branch of knowledge that's largely neglected by modern scholars, who tend to view it as being little more than a collection of superstitions and gross errors about the nature of chemical reactions. Alchemical writings are often bizarre, reading more like myths than scientific papers. Many have been lost, and those that remain are difficult to interpret because alchemists made their treatises deliberately obscure. They used code words, metaphors, and puzzles to reveal their meaning only to fellow alchemists.

Alchemy can be divided into three overlapping branches: practical alchemy, chrysopoeia (the transmutation of base metals such as lead into gold), and philosophical alchemy. Alchemists in the first group adopted an approach similar to the one modern chemists take, discovering methods for producing valuable commodities such as inks, dyes, alloys, and perfumes.

The second group of alchemists, those who pursued the art of chrysopoeia, is the most famous. Many were honest students of the chemical arts, but this branch of alchemy also attracted charlatans whose main occupation was cheating the wealthy. Throughout its history, this branch cast a shadow of fraud over the practice of alchemy, which sometimes drove its devotees underground to experiment in secret.

The belief that lead could be transformed into gold may seem bizarre, but the theory behind it makes a certain amount of intuitive sense, and is not so different from the theory on which modern physical science is based. Alchemists believed that all substances consist of two components: form and matter. Matter makes things corporeal; form gives things their specific properties, such as color, shape, and texture. The underlying matter is the same for all things, whereas the form is what makes each thing unique. You need both matter and form for a thing to exist. You can't have matter by itself, because you can't have a substance that has no properties: no shape, texture, color, or weight. Pure form can't exist on its own either. It's incorporeal; it has no body. It's just an idea. Existence requires the combination of both.

Although form and matter are equivalent in the sense that both are necessary for existence, they're different when it comes to our ability to understand them. Consciousness can comprehend form because it's used to working with abstract ideas. Undifferentiated matter, called *prime matter* by the ancient philosophers, is different. The ego can't comprehend it because the ego can't imagine something that has no properties. If you try, you might imagine a black blob—but black is a color and a blob is a shape, both of which are properties that give matter form. To the ego, matter is a deeper mystery than form. We get a clue to the reason for this mystery when we note that the word *matter* is derived from the Latin *mater*, which means "mother." Matter is a manifestation of the great mother archetype, which operates from deep within the collective unconscious.

How did the theory of form and matter translate into the belief that lead could be turned into gold? Since all things are made of the same stuff—prime matter—it should be possible to strip lead's matter of its base form (base in the sense of ignoble) and replace it with the noble form of gold. The alchemists looked for ways to separate the "spirit" of a substance (its form) from its body. By then combining that spirit with another body, they hoped to bring about the transmutation of one substance into another. Think of the transformation of elemental carbon into diamonds. They're both made from the same substance, but their forms are so different that one is worthless, while the other is a valuable gem. If you heat a diamond in the absence of oxygen, you can drive off its form (in the alchemical sense) and turn it back into elemental carbon.

Lead and gold may not be made from the same substance the way elemental carbon and diamonds are, but the idea that they share the same underlying matter is still accepted as scientific fact. Modern chemists believe that all things are made of protons, electrons, and neutrons. In theory, you could turn lead into gold by changing the number of protons in its atomic nucleus. Lead has eighty-two and gold has seventy-nine. The reason the alchemists were unable to bring about this transformation is that the number of protons in a nucleus can't be changed using chemical reactions.

Understanding the world by distinguishing form from matter is also a recurring theme in other magical traditions, which brings us to the third branch of alchemy, the one most relevant to the study of magic and the unconscious. It's called philosophical alchemy, because its goal was to probe the secrets of nature by studying alchemical phenomena. The aim of this group of alchemists was to discover the secret of transformation of the base into the noble, the transitory into the eternal, not only in metals but in the world at large. They sought to unlock the secrets of physical matter so they could mingle reason with nature and thereby ennoble the world through mystical transformation.

It's difficult to know precisely what they actually had in mind. Like most aspects of alchemy, the process—even the goal—was shrouded in mystery. Nevertheless, it's not hard to imagine why these alchemists saw chemical reactions as a fitting symbol for transformation. When two chemicals react with each other, something occurs that's different from what we're used to seeing. Usually, when we combine two things, we get a mixture that has some of the properties of each, like when we pour cream into coffee. A chemical reaction, on the other hand, creates something new, something that may be radically different from the parent compounds. Basil Valentine, a fifteenth-century alchemical philosopher, described a process by which metallic gray antimony is transformed into a yellow, transparent glass.[1] An internet search for "chemistry magic" returns dozens of entertaining videos in which liquids change color, fire erupts from ice, and solid objects rise up into the air.

The philosophical alchemists based their theories of transformation on the work of the fifth-century BCE philosopher Empedocles, who proposed that chemical reactions occurred under the influence of love and strife. Love was responsible for two substances coming together to form something new (a personification not so different from the modern convention of describing certain chemicals as having an affinity for one another). Strife, by contrast, was responsible for the breaking apart of substances into their component parts. Love and strife were also called *eros* and *logos*. *Eros* was the Greek god of passionate desire. *Logos* gives us the English word *logic*, and the suffix *-logy*, which we use to designate the study of something, like "biology." *Logos* refers to analytical thinking.

Eros and *logos* not only personify chemical reactions, they also give us another pair of symbols for the unconscious and the ego. *Eros*—passionate desire—is related to the unconscious. It's irrational, emotional, affects both the body and the mind, and operates outside of voluntary control. It also reflects the joining behavior of the unconscious, which connects things using the laws of sympathetic magic. *Logos*—analysis—is related to the conscious mind. It's rational, under voluntary control, and it's the favored tool the conscious mind uses to understand the world. Jung described *eros* as "psychic relatedness" and *logos* as "objective interest."[2] *Eros* is associated with intuition, acceptance, and simple being. *Logos* is associated with productivity, goal orientation, intellectualism, and making an impact on the world.

If we think back to Walt Whitman's mystical understanding of the stars as he gazed on them in silent appreciation, we might see that understanding as a kind of connection that could be described as *eros*. When we analyze something, we keep it at arm's length. Our object of study is clearly separate from ourselves. But when we appreciate something without analyzing it, the separation becomes less distinct. The stars in the sky take up residence within the mind as an impression. The impression isn't dissected with analysis, but rather allowed to remain whole—an inner reproduction of the original—such that the stars in the sky and the impression in the mind are joined by the sympathetic law of similarity.

You may recall from an earlier chapter that Jung believed that traditions of magic, and the narratives they're based on, are descriptions of the unconscious. The work of the philosophical alchemists was of particular interest to Jung because he understood their work of chemical transformation as a metaphor for individuation, the psychological transformation brought about by the transcendent union of the ego and the unconscious.

His interest began when he happened to open an old book on alchemy one day, and found to his surprise many of the same themes and symbols his patients had described in dreams—ones that neither Jung nor his patients had any familiarity with. He began avidly seeking out alchemical manuscripts, both well-known and rare.

One alchemical manuscript, Basil Valentine's *Twelve Keys*, provides an example of how alchemists used symbolism to describe their chemical reactions:

> Take the ravenous grey wolf that on account of his name is subjected to bellicose Mars, but by birth is a child of old Saturn, and that lives in the valleys and mountains of the world and is possessed of great hunger. Throw the king's body before him that he may have his nourishment from it. And when he has devoured the king, then make a great fire and throw the wolf into it so that he burns up entirely; thus will the king be redeemed.

Lawrence Principe, a professor of chemistry at Johns Hopkins University, interprets this passage as a method of purifying gold.[3] Gold is the metaphorical king of metals. The wolf is stibnite, a gray mineral made from sulfur and antimony that Principe suggests is "ravenous" because, when it's melted, it dissolves other metals "with breathtaking speed." When impure gold is added to molten stibnite, the impurities combine with the sulfur and float to the surface. The gold forms a brilliant white alloy with the antimony, and when this alloy is heated over a fire, the antimony is driven off, leaving behind pure gold, the "redeemed king" in Valentine's description.

Principe believes the personification of chemical reactants was simply a way of hiding the alchemical secrets from those who didn't have the background necessary to decode the riddles. Jung, however, interpreted alchemical writings as expressions of archetypal themes. In just the brief quoted passage, we find the archetypes of royalty, the predatory animal, death, and redemption by fire.

An alchemical emblem illustrating, in symbolic form, the purification of gold. The old man with the sickle is Saturn, who was associated with lead. Like the ravenous wolf (stibnite), he's being burned in the fire because lead is a common impurity found in mined gold. ("The First Key of Basil Valentine," 1678.) *(Source: Wellcome Images.)*

Jung saw these themes as manifestations of the collective unconscious. They were prominent in alchemy because alchemists projected their unconscious ideas onto the chemical work they were doing. This projection made the art of alchemy an exploration of unconscious contents and alchemical recipes a representation of the labor that the ego, as metaphorical alchemist, has to undergo to make the unconscious conscious.

PROJECTION REDUX

Earlier, we encountered projection when we looked at hypocrisy and denial as consequences of projecting one's shadow onto another person. That's a specific (and dangerous) form of projection, but projection can also be understood in a broader sense: as all the ways a person modifies their perception of reality to mold it into a reflection of her unconscious. Or as Jung writes, "Projections change the world into the replica of one's own unknown face."

To understand how projections work, consider the simple example of an optical illusion.

The gray square illusion. Despite appearances, squares A and B are the same shade, as revealed by the connecting bar. ***(Source: Edward H. Adelson via Wikimedia Commons.)***

The conscious mind, as we've discussed, receives a highly processed version of the data coming from the sense organs. Before the signals are consciously experienced, unconscious brain circuits translate them into patterns the conscious areas of the brain can work with. In the process, those unconscious circuits impose their own interpretation on the data. This translation and interpretation creates a projection. The unconscious decides what the sensory data means, and that—not an accurate reproduction of reality—is what you perceive.

Look at the figure on the left. If you're like most people, you perceive square A as darker than square B. The difference between the two seems beyond doubt. But the picture next to it shows that, in fact, they are the same shade. The projection is so compelling, however, that even with the help of the connecting bar, the two squares may still appear different. You need to cover the rest of the picture with your hands to see they're identical.

The reason square B looks lighter is because it's in the shadow of the cylinder. Things that are in shadow appear darker than they really are, so the brain makes a correction. The brain assumes that because the square is in a shadow, it must really be brighter than its luminance indicates.

In classical Greek philosophy, the word for this kind of subjective sense experience was *phantasiai*. It was understood as raw data coming from the sense organs blended with judgments about its meaning. Thus, the *phantasiai* of square B in the figure is the shade of gray the eyes register, "corrected" by the judgment that its actual color must be lighter because its luminance is reduced by the shadow of the cylinder. The word *phantasiai*, of course, is linguistically related to "phantom" and "fantasy," experiences that have no basis in reality, reminding us that what we perceive and what's really out there may be different things. Much of what we believe to be objective reality is actually a reflection of unconscious preconceptions. Sometimes the difference is just a minor correction in luminance. At the other end of the spectrum, a *phantasiai* can invest the world with the power of the archetypes.

If you take a piece of cardboard cut in the shape of a V, and shine a light on it from above, it casts a V-shaped shadow. If you move that shadow across a baby chick from side to side, nothing happens. But if you move it from back to front, the moving shadow activates a projection of the predator archetype (a hawk), and the chick becomes frantic. Some projections modify our world in subtle ways, creating an optical illusion or giving natural phenomena emotional meaning, such as when a cloudy sky appears gloomy. But projections that activate the archetypes can lead to terror, ecstasy, even a complete reevaluation of one's life.

Some archetypes can be activated in predictable ways. You can stimulate the predator archetype in your friend with a rubber snake, for example. But other archetypes, those that reflect the unique characteristics of an individual's psyche, are triggered in less obvious ways. In general, things that are unfamiliar are more likely to invite projections. If something is unfamiliar, your brain has to make more assumptions about what's being perceived, and who you are as an individual determines what you project. Anxious people tend to project danger onto their surroundings, and so they often find unfamiliar environments and people threatening. Highly social individuals, by contrast, project benevolence onto others, which is why they enjoy meeting new people.

Things that are indistinct also invite projection because the brain fills in the missing pieces. When you go outside after a snowstorm and see the trees covered with a blanket of snow, it seems more magical than usual. The snow hides the familiar, encouraging archetypal projection that infuses the scene with magic. A dark room is creepier than a well-lit one for the same reason. When you go to a monster movie, you usually won't get a good look at the monster until the end. The director knows that the monster you can't see is scarier than the one you can, because it invites projection from your unconscious—the same place nightmares come from—that no special effects can match. Moonlight and candlelight have the opposite effect. They make a scene more inviting by encouraging projections of mystery and romance.

Projections can induce non-ordinary states of consciousness. As you gaze at that first snowfall of winter, William James's filmy screen of the ordinary is broken through, and, for a few moments, a spell of enchantment transforms the world around you. Something similar happens in the dark theater as a shadowy threat darts across the screen. The world feels different compared to how we feel commuting to work or picking up a quart of milk at the grocery store.

A dark, creepy room or an enchanted snowy day lifts us out of the commonplace routine of daily life, shifting us into a subtly altered state of awareness. More dramatically, projection of the archetypal image of the ideal masculine or feminine can make ordinary humans appear as gods and goddesses. This projection is responsible for the temporary insanity we call "falling in love," which can generate a non-ordinary state of consciousness that lasts for months. Projection can charge nearly anything with the vitality that comes from the unconscious. Jungian analyst Marie Louise von Franz describes the effects of projection on the learning process:

> If you have to learn about a subject you do not love, where you have no projection, which means that you have no relationship to it, it does not mean anything to you and is not connected with your flow of libido [psychological energy from the unconscious], so you have to toil and sweat at something and learn it for the exam, but ten minutes later you have forgotten it again. If you are fascinated, however, which means that projection has taken place, then you get

> emotional and acquire a tremendous amount of consciousness very easily and quickly. That is the whole secret of teaching and learning. Fascination always involves projection.[4]

Unlike the projection of the shadow, which is always bad, everyday projection has both advantages and disadvantages. By producing the emotional connection described as *eros*, it gives the things around us the power to capture our interest—a clear benefit. And by letting us encounter our unconscious in the outside world, projection gives us the opportunity to integrate a portion of its contents, moving us forward along the path of individuation. On the negative side of the ledger, projection distorts objects and people in our environment. We don't see what is actually present, and instead see (to use Jung's formulation) a mirror of our own hidden face. That distortion may lead to bad decisions and unnecessary conflict with the people around us, like the anxious person who projects unfriendliness onto strangers, an example of the unconscious connection forged by *eros* producing something negative.

It may seem odd to use a word typically associated with love and desire to describe projection of negative contents. But in Jungian psychology, *eros* is used to describe any emotional connection that isn't consciously chosen, including negative ones. For example, when we feel antipathy for someone, he gains power over us—the power to annoy. He may do it intentionally or accidently; unconscious agents may do it without him being aware of it. He occupies our thoughts more than those we're indifferent to, and we have stronger emotional reactions to the things that happen to him. A bond—*eros*—has been formed, and that bond is difficult to break.

CORRESPONDENCES

Projection creates connections not only between people but also between things found in the natural world and psychological experiences. When we project archetypes of majesty and power onto thunderbolts, the inside and the outside become linked. Fanciful connections of this nature are common. The law of similarity makes gold a king and stibnite a wolf.

In alchemy, the law of similarity is called "correspondence," and it's most clearly expressed in the maxim "As above, so below," a condensed version of a line from the Emerald Tablet of Hermes Trismegistus, the most famous of the alchemical texts. The original reads, "That which is Below corresponds to that which is Above, and that which is Above corresponds to that which is Below, to accomplish the miracle of the One Thing." Its primary interpretation is that the things that happen in the macrocosm (the universe) correspond with the things that happen in the microcosm (an individual person) and vice versa. We see this most clearly in astrology, in which the positions of the stars and planets at a person's birth are said to influence his personality and fate in life. The meaning of "the One Thing" is obscure in this passage, but the concept of the One is important in alchemy. The mystical aphorism "all is One" is the basis of correspondences. If all of existence flows from one great unity, then everything is linked in some hidden way.

We've seen how all archetypes have a light side and a dark side. The gods both bless and curse. Another duality that finds unity within the archetypes is the chthonic and the spiritual. "Chthonic" refers to things that are in or under the earth. Symbols of chthonic archetypes include caves, creatures who live underground, and terrestrial nature in general. Archetypes that are oriented upward are spiritual and may be symbolized by birds, stars, or angels.

We can look at Venus, the goddess who personifies the archetype of love and desire, as an example. Her influence—the correspondence that links the planetary deity with human experience—has both chthonic and spiritual elements. One of the chthonic aspects of Venus is physical desire, whereas a spiritual aspect might be selfless love that puts the welfare of the beloved above one's own. The ego may judge Venus's spiritual side to be the superior one, but both are essential. Without the chthonic side, no egos would be born to pass judgment. The unification of the spiritual and the chthonic within the archetypes is another manifestation of "As above, so below." The higher and the lower are one.

THE FOUR ELEMENTS

At first glance, the chemical theories of the alchemists seem old and outdated. But a closer examination finds that they permeate our culture, even the language we speak. Since the alchemists' theories are based on intuitive psychological truths, they're timeless, and if we suspend disbelief and try to understand them more as stories than as scientific theories, we'll be able to appreciate their lasting value. By examining three of these theories in particular—the four classical elements, which alchemists believed were the basic building blocks of all matter; the planets and their associated metals; and two important alchemical processes, the *circulatio* and the *conjunctio*—we'll also gain an expanded understanding of unconscious modes of thought.

We'll begin by looking at the four elements: earth, water, air, and fire. The theory that all things are made from these four elements is thought to have originated with Empedocles. The elements roughly correspond to the four natural states of matter in modern chemistry: solid, liquid, gas, and plasma. Unlike the states of matter that modern chemists understand, however, each of the four elements had symbolic meaning derived from philosophical speculations—or, we might say, projections of unconscious contents.

Earth was associated with the qualities of heaviness, stability, and a downward orientation, because heavy things fall. It was also associated with the mother archetype (as in Mother Earth), particularly with regard to fertility. The physical body and the senses we use to perceive the material world were linked to earth as well.

Water represented all kinds of fluids. It was considered a receptive element; because it engulfs and holds, it was linked, like earth, to the mother archetype. Water was also associated with emotion because of its fluid nature.

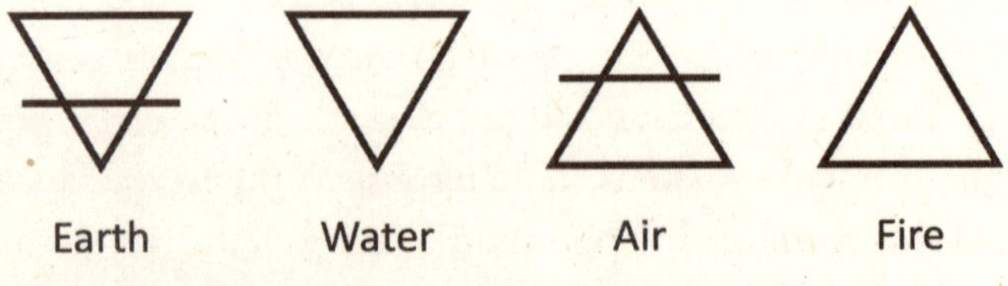

The alchemical symbols of the four elements.

In English, the connection between water and emotion can be found in a number of idioms that portray a person as a container for fluid emotions, including "She was filled with emotion," "He poured out his feelings to her," and "He bottled up his emotions."[5] We also speak of "a surge of anxiety," "a gush of relief," and "a flood of emotion."[6] Water is where life began, and it is essential for all living things, so water also represented life. In fairy tales and myths, we sometimes encounter a magical substance called "the water of life" that brings healing and renewal. One example is the mythical Fountain of Youth.

As we continue upward, the next element we encounter is air. Air was associated with the intellect and spirit, which are rarefied and insubstantial. Because air has no fixed shape, it can fill or surround other objects, reflecting the intellect's ability to accommodate an unlimited variety of ideas. Air is also able to penetrate some objects to get at their inner essence. Like the intellect, which can jump from one idea to another, air flies from place to place in the form of wind, a symbol of the spirit. The Hebrew word *ruach* can mean "wind," "breath," "spirit," or "mind." The English word *psyche* is derived from the Greek *psūkhein*, which means "to breathe," and in some religious traditions, humans were awakened into consciousness by the breath of the creating deity blown into their nostrils.

The last element is fire, which was associated with motion, change, and, like air, an upward orientation. In alchemy, heat from fire was often used to initiate chemical reactions; thus fire was associated with change and transformation. Fire was similarly associated with the spark of inspiration—new ideas that change the course of a person's life—and the passion needed to convert mental inspiration into something real that impacts the external world. Transformation involves the death of the old and the birth of the new. It's both destructive and creative. When gray, metallic sodium reacts with yellow-green chlorine gas, both pass away, and white table salt is created. When William Wordsworth wrote "the Child is the father of the Man," he was describing something similar. As we go through life, we undergo countless transformations in which our old self dies and a new self is born. Those who resist these transformations, holding fast to things that have become worn out, place themselves in opposition to this rule of life. Sometimes the transcendent function brings about a catastrophe that destroys everything an individual holds precious in order to force them back on the road to growth. When things have gotten to the point where they can't be fixed, we say, "Burn it all down and start over." In mythology, this process is symbolized by the phoenix, which grows old and decrepit, then bursts into flame and is reborn as a hatchling. We see a reflection of this process in the alchemical recipe for purifying gold. The king, who has been devoured by the wolf, is reborn and purified through the agency of fire.

Whereas air represents the thinking aspect of the soul, fire represents its animating aspect. The soul not only cogitates; it is also the spark that turns inanimate matter into a thing that moves, a thing that pursues goals and desires. Fire is energy and action.

The four elements, or states of matter as we call them today, are fundamental to our experience of the physical world. Consequently, expressions that use the states of matter metaphorically are common. We have a "solid performance," a "fluent speaker," an "effervescent personality," and an "inflammatory statement." Chemistry is the poetry of everyday life. Each state has its own personality, an alchemical reaction within the human mind between the physical and the psychological.

Projection animates the world by endowing physical objects with personalities. It makes unconscious agents visible and easier to understand. Perhaps the most elaborate personalities that have been projected from the unconscious are the gods and goddesses of ancient myths. Their divine nature raised them far above mortals, but their influence wasn't confined to the heavens. Their spirits inhabited not only planets and constellations, but also earthly matter, most especially metals, and that placed them within reach of the alchemists. They put the gods and goddesses in vessels and flasks; they melted them, boiled them, and prayed over them in the hopes of unlocking their mysteries and harnessing their power.

THE PLANETS AND THEIR METALS

The word *planet* comes from the Greek *planasthai*, which means "to wander." Planets were called wanderers because they were constantly changing their position relative to the background stars. Although stars appear to revolve around the earth as the night progresses, they don't move relative to one another, so the ancients referred to them as the "fixed stars" (i.e., fixed in place). The stars and the planets were thought to be attached to enormous transparent spheres that rotated around the earth. The sphere of the fixed stars was the outermost one.

The planets of the ancients were those that are visible to the naked eye: the Moon, Mercury, Venus, the Sun, Mars, Jupiter, and Saturn, named after the gods who, they believed, held power over the minds of mortals. Although modern astronomers don't classify the Sun and Moon as planets, they met the ancient definition because they move in the sky relative to the fixed stars, rather than staying in one place. The Sun, for example, moves from north to south and back again as we progress through the seasons of the year.

The order of the planets was determined by how long it took each one to make a complete circuit through the heavens, which, except for the Moon and the Sun, is approximately the length of time it takes a planet to orbit the Sun. The ancients didn't know that space was a near vacuum, so they assumed that the planets and the stars made some sort of sound as they whizzed through the air. Each of the spheres was thought to produce its own unique tone, and the combination of these sounds was called the music of the spheres. The reason we can't hear the heavenly music, they theorized, is because we're so used to it. It's present from the moment of our birth, and so, like the hum of an air conditioner, it fades into the background, unnoticed.

Each of the planets was associated with a god or goddess who gave the planet its unique characteristics. Each planet was also associated with a terrestrial metal that shared the planet's qualities. Copper, the metal of Venus, is relatively soft and has a pinkish-orange hue reminiscent of tanned skin, whereas iron, the metal of Mars, was particularly well suited for crafting

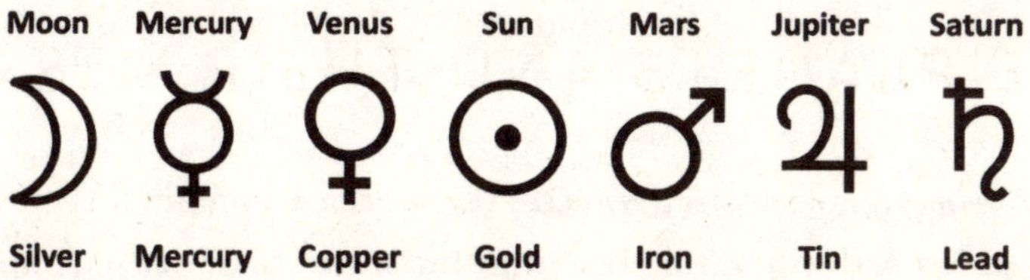

The alchemical symbols for the classical planets and their metals.

weapons of war. This correspondence of planets and metals is a manifestation of the maxim "As above, so below." The alchemists believed that a single spirit infused both. The qualities that characterized the planet/divinity were also found in the metal.

The planets not only determined the properties of terrestrial metals, but they also influenced the human psyche with their distinctive characteristics. From a psychological perspective, the planetary gods are personifications of agents within the unconscious, and like all unconscious contents, they have a light side and a dark side; they help and they harm, via emotions, passions, and desires, among many other things. But these personifications are more than conglomerations of attributes. Even when they contain their own opposites, they give the impression of coherence. Mars, for example, inspired both unrestrained fury as well as martial discipline; the god of war encompassed both. Royal Jupiter, the king of the gods, was the patron of justice, who punished those who broke the trust of sacred oaths. But he was also the jovial god of festivity, even of excess and overindulgence. The gods are archetypal, making them more than the sum of their parts.

A description of each of the planetary gods can be found in the appendix. Here, we'll look only at two of them, Mercury and the Sun, each of which represents archetypal aspects of consciousness. (Archetypes are typically associated with the unconscious, not consciousness; however, as part of the psychological bedrock all humans share, consciousness also has its own archetypal blueprint.)

Mercury names the planet, the god, and the metal, which are all associated with one another. Swift, cunning, eloquent, and diplomatic, Mercury was the god of adaptability, variability, the intellect, and language. The metal mercury is also known as quicksilver, because it's a liquid that shines like silver. Like other liquids, it can take on any shape, reflecting the power of language to represent almost anything that exists. When the metal mercury is disturbed, it breaks apart into a multitude of droplets, then instantly reforms. The animated metal's abilities to reflect things off its surface, and to break apart and recombine in numerous ways make it an apt symbol for the intellect and for language, which expresses the intellect's ideas.

Because of his dominion over things that are broken apart and recombined, Mercury was the patron god of the alchemists. The ancient Greeks

Mercury, the winged messenger of the gods.
(Source: Uva79 via Wikimedia Commons.)

called this god "Hermes," and alchemy is sometimes called the "Hermetic art."

Language can be both productive and destructive. The metal mercury shines with reflected light, but it's also slippery, evasive, and toxic. Language can reveal truths, but it can also reflect things that don't exist, such as stories and lies; thus Mercury also played the role of the trickster, and was the patron god of thieves. One of the evolutionary advantages of consciousness is that it gave physically weak primates the ability to outsmart nature, using tools such as fire, rope, levers, and other technologies. Modern farming techniques allow humans to grow food in the desert. We burn oil to warm our homes in the winter. Electricity gives us light when the sun is down. Like a trickster, the conscious ego ferrets out hidden knowledge and uses it to break the natural order for its own advantage. Like a thief, it takes what it wants from the Great Mother, whether she is willing or not.

The Sun, personified by the Roman god Sol, symbolizes logic and rational thought. The Sun's metal is incorruptible gold. Whereas the Moon is associated with changeable emotion, the Sun symbolizes unchanging, eternal reason. In the Greek pantheon, the god associated with the Sun was Apollo.

Robert Fludd's "The Sun," from *Utriusque Cosmi*, 1617.
(Source: Journal of World-Systems Research, *vol. 21, no. 1.)*

His principal shrine was located at Delphi, and carved into the entrance of the temple were the words, "Know thyself," perhaps the most famous aphorism in all of philosophy. For the Greek philosopher Socrates, "Know thyself" was the guiding principle of his life. He claimed to know little about most topics because, he said, it would be ridiculous to study other things when he did not yet understand his own self. Similarly, Jung believed that attaining full knowledge of the self was the primary task of the ego. The sun of consciousness must shine its light into the primordial darkness of the mind.

The alchemists understood their chemical reactions as terrestrial manifestations of a celestial drama of love, conflict, and divine transformation. They believed the metals they worked with were inhabited by the spirits of the planetary deities who gave them their virtues, such as strength and luster, as well as their vices, such as corruptibility and toxicity. From a Jungian perspective, the alchemical gods are the archetypes of the collective unconscious, exerting their influence from inside the mind. The stories told about these gods and goddesses tell us about ourselves and the forces that drive us from within. The unconscious is not one homogeneous mass. There are characters in there, characters that have personalities, great power, and overwhelming beauty—but they're dangerous. The ancient myths about gods and goddesses don't sugarcoat it. They tell stories of matricide, patricide, infanticide, cannibalism, incest, torture, and many other horrible topics. But they're fascinating.

THE *CIRCULATIO* AND THE *CONJUNCTIO*: LABOR AND LOVE

The alchemists were familiar with many of the methods that chemists use today, including distillation, fermentation, and the use of solvents. These practical tools were supplemented by processes that were more philosophical in nature. Jung was particularly interested in two of these alchemical processes: the *circulatio* and the *conjunctio*.

The *circulatio* was the repeated processing of alchemical ingredients, viewed as necessary to bring about the desired results. Here's how von Franz described it, and its relation to psychological transformation:

> The *prima materia* has to be constantly washed and distilled and thus the first activity of the opus is distilling, washing, and cleansing, over and over again. Here it says nine times, others say fifteen times, and some say ten years. It is really a very long process and sometimes means endlessly rehearsing the same problem in its different aspects. That is why also in alchemical texts they always allude to the fact that this part can go on for a long time and is characterized by endless repetitions just as, unfortunately, we fall again and again into complexes which have not been worked out and have to be looked at time and again.[7]

The kind of growth that leads to transformation requires many failed attempts. Quitting smoking, for example, requires an average of ten attempts before long-term success is achieved. As Mark Twain said, "Giving up smoking is the easiest thing in the world. I know because I've done it thousands of times." Thomas Edison experienced the challenges of the *circulatio* as he developed the incandescent light bulb. It was nearly impossible to find a filament that wouldn't immediately burn out. But Edison persisted, working his way through the trials of the *circulatio*. He said, "I have not failed. I've just found 10,000 ways that won't work."

The need to repeat the same challenges and make the same mistakes over and over before we can achieve lasting growth is a common theme in religion. Belief in reincarnation is prominent in Buddhism and Hinduism, and was also shared by ancient Greeks, Romans, Celts, and Teutons. While Christian theology doesn't support reincarnation, Catholics believe in purgatory, a place to be purified prior to entering heaven. In the *Divine Comedy*, Dante depicts purgatory as a mountain that souls climb by traveling a circular path, overcoming one of the seven deadly sins with each revolution.

Some cathedrals contain a labyrinth within the nave, symbolizing the journey out of sin toward redemption. At first, as you enter the labyrinth, it appears that you're headed straight for the center. When we first confront

Dante's purgatory is reminiscent of the *circulatio*. At the summit we find a symbol of the *conjunctio*, the unification of complementary opposites. (Domenico di Michelino's *La Divina Commedia di Dante*, 1465.) *(Source: Wikimedia Commons.)*

Grace Cathedral interior with labyrinth.
(Source: David Clay via Wikimedia Commons.)

a problem, it often seems like we can solve it by simply making a rational choice to change our behavior. In most cases, rapid initial progress soon gives way to setbacks, and just like in real life, the pilgrim walking the labyrinth finds that the path first leads her in a direction away from her goal. There are no shortcuts. To reach the center, all four quadrants of the labyrinth must be traversed.

Once we recognize the *circulatio* in our own lives, we can see it in others' lives, as well—a recognition so important, it's a moral imperative within the Christian religion. In the Gospel of Matthew, Peter asks Jesus, "Lord, how many times shall I forgive my brother or sister who sins against me? Up to seven times?" Jesus answered, "I tell you, not seven times, but seventy-seven times."

We saw a fairy tale example of the *circulatio* in "The Golden Bird." The young prince disobeyed the fox repeatedly, getting himself sentenced to death each time. The fox was annoyed but patient, never forsaking the prince, always ready to give him another chance. A modern story about the *circulatio* is the 1993 Bill Murray comedy *Groundhog Day*, where a selfish, cynical man must repeat the same day over and over until he gets it right by learning the lessons of compassion and love.

The *circulatio* often refers to a person's inner efforts to achieve growth, but it can also apply to the world we live in. This aspect of the *circulatio* is symbolized by the legendary Wheel of Fortune, which never stops turning, arbitrarily bringing good or ill luck to us mortals. The wheel is beyond our control, and the only thing we can do is accept its revolutions with equanimity. This kind of tranquil acceptance of the ups and downs of life plays a central role in Buddhism and in the philosophy of the Stoics.

The *circulatio* is ultimately followed by the *conjunctio*, the bringing together of opposites, which for Jung was a symbol of individuation. In alchemy, some reactants were considered masculine, while others were considered feminine—although the identities were inconsistent, because in one context a substance could be masculine and in another feminine. Acids were usually masculine because they attack and corrode other substances in a seemingly warlike, aggressive way. Silver, by contrast, was considered feminine because of its smoothness and beauty. Its tendency to tarnish reflected its changeable nature, which was at the time considered characteristic of emotional femininity, in contrast to unchanging rational masculinity.

The *conjunctio* was the joining of the two opposites. In alchemical illustrations, it was symbolized by sex, marriage, and the hermaphrodite, a formidable creature who was both male and female.

Many mythological traditions portray the original humans as hermaphrodites, representing a state of wholeness. In Plato's *Symposium*, a dialogue about the nature of love, the poet Aristophanes relates to his audience a myth about the primeval human who was both man and woman. Together, they were so powerful that they climbed up to heaven and made an attack on the gods. In order to defend the gods and put humans in their place, Zeus

Hermaphrodites had great power because they possessed completeness. The Y represents the essential being of nature, which is both male and female. Although the hermaphrodite is a creature from pagan mythology, the presence of the Pope in this illustration is intended to bring the hermaphrodite into the fold of Christianity. (Engraving from Michael Maier's *Symbola Aureae Mensae*, 1617.) *(Source: Wikimedia Commons.)*

cut the hermaphrodites in half. Aristophanes claimed that romantic love is the yearning these two halves feel for each other, a reflection of their desire to regain this original state of wholeness.

In Hindu mythology, the god Ardhanarishvara is composed of both the male Shiva and the female Parvati, representing the synthesis of the masculine and feminine energies of the universe. This synthesis takes us back to the ouroboros because their divine union generates the entire universe and represents the eternal reproductive power of nature. When the two sexes are divided, this ouroboric power of self-fertility is lost.

The word *hermaphrodite* is a portmanteau combining the names of two Greek gods: Hermes and Aphrodite, corresponding to the Roman Mercury and Venus. Mercury represented *logos*, or objective, analytical thinking;

whereas Venus represented *eros*, the subjective connection to the world driven by fascination and desire.

In alchemical recipes, the *conjunctio* produced a mystical, divine water called the "bitter water." It's bitter because the initial contact with the unconscious is unpleasant. When we turn our attention to the unconscious, the first thing we encounter is our shadow. Similarly, when the princess first met the frog, she found it disgusting. At the beginning of individuation, the insight we gain into ourselves is not always flattering. We must face "the bitter truth," which is not always comfortable and can discourage us from pursuing self-knowledge. But if we persevere, we reach the ultimate goal—symbolized in alchemy by the philosophers' stone.

THE FINAL DESTINATION

The endeavor to make the philosophers' stone was called the *Magnum Opus*, the Great Work. It was the stone (actually described as a liquid or powder) that had the ability to make base metals noble. It was the elixir of life, which led to rejuvenation, and symbolized perfection, enlightenment, and bliss. Jung believed the philosophers' stone was a symbol of the self.

A stone isn't an obvious choice to represent the goal of a lifetime of work. Stones are generally things of little value that we look down on. A stone lies on the ground where we step on it, kick it. It does nothing except lie there. Why would something so lowly symbolize the highest goal? Von Franz saw a clue in the artistic use of stone to represent eternal permanence, for example in monuments and grave markers. Similarly, the eternal word of God in the form of the Ten Commandments was written on stone tablets. Another notable use of stone in the Bible is found in the Book of Revelation. God is quoted as saying he will give a white stone to those who are steadfast, and "on the white stone is written a new name that no one knows except the one who receives it." Here, the symbol of the stone does double duty, representing both eternity and individuality.

The motionless stone can also represent tranquility in the midst of life's turmoil. Von Franz notes that sometimes, after a great deal of struggle and suffering, something breaks through. A quiet tranquility replaces our frantic efforts to solve an insoluble problem.

Psychiatrists see this often with substance users who are trying to overcome their illnesses. Substance users who seek treatment are usually ambivalent. They do so because they recognize that their lives are going in the wrong direction. They know the drug is harming them and something needs to change, yet they don't want to lose the good parts of substance use. They often have little left besides the drug. After years of use, other interests are gone; their family life is in shambles, the drug is their only comfort, and they can't imagine life without it. As a colleague's patient once asked her, "What if someone told you that you had to give up your career as a doctor? That's how it feels for me to think about giving up cocaine." They struggle to find a way to keep the pleasures of the drug while putting their lives back together. But sooner or later, they realize it isn't working. Their life will have to change in almost unimaginable ways. And yet, unexpectedly, they're at peace. When the thing they feared most becomes a reality, they find tranquility. They become the immovable stone. Addiction specialists are taught to watch for the emergence of this stillness. It marks a turning point, when the therapy needs to transition from discussion to action.

The starting point of the Great Work is prime matter, an inherently base and contemptible substance. The material used to make the philosophers' stone "is of cheap price and found everywhere," according to one ancient alchemical axiom. "It is found in filth" runs another.[8] The alchemists weren't so different from the princess in "The Frog Prince." The initial contact with the primeval unconscious disgusts the ego, but through labor and perseverance, what first appears ugly can be transformed into something of great value. For the princess, it was her true love; for the alchemists, it was the philosophers' stone.

It's not easy to overcome one's initial aversion to the primitive aspects of the psyche, but that's where the work of individuation begins—the work of making the unconscious conscious so that it can be raised up into the light and humanized. Jung quotes the alchemist Morienus: "Take that which is trodden underfoot in the dunghill, for if thou dost not, thou wilt fall on thine head when thou wouldst climb without steps."[9] If you reject what you consider to be low and contemptible, you will lose the base you need to stand on, as you reach toward your higher self.

Human beings are alchemists who practice the art upon themselves. But our approach is not the same as the ancients'. Psychology teaches us that our projections paint the world with the colors of the unconscious. The planetary gods who control the fate of humanity have taken up residence within the human psyche, and the cosmic alchemical maxim "As above, so below" becomes personal: As within, so without.

The fairy tales we looked at in the previous chapter focused on the chthonic aspects of the archetypes, our connection to the earth, and our instinctual, animal nature. Alchemy carried us upward by connecting the chthonic spirits that inhabit the metals of the earth with the planetary gods. In the next chapter, we continue our journey upward, traveling beyond the earthy, chthonic manifestations of the archetypes—away from animals and metals—to the rarefied abstractions of numbers.

MYSTICAL NUMBERS

Mathematics is the language with which God has written the universe.

GALILEO GALILEI

PHILOSOPHY AND INTUITION

Numbers may seem sterile and dry compared to the archetypal narratives found in fairy tales and alchemical lore. But the cosmic progression of the mystical integers reveals a story as powerful as any myth. It tells the story of individuation, carrying us from the undifferentiated state of oneness with nature to the mystical state of full integration of all aspects of the self.

Like fairy tales, the story of mystical numbers goes back thousands of years and has evolved over time. But it's a different kind of story. While fairy tales appeal mainly to emotion, numbers acquired their mystical meaning from philosophy, which is based on reason.

When a fairy tale hits home and we know it's telling us something important, it *feels* right. Philosophy, by contrast, doesn't care much for feelings, preferring rational deliberation. But even the most rational philosophy is based on irrational intuition, a product of the unconscious. As the alchemist Morienus reminds us, the higher cannot stand without the lower.

It's important to keep in mind that irrational is not the same as wrong. An irrational belief is simply one that isn't based on rational thought processes. Irrational thinking can have great value because not all problems can be solved with a rational approach. Additionally, as we saw in a previous chapter, the nonrational parts of the mind have a great deal of information-processing capacity at their disposal that can help us manage problems too complex for conscious processing. In many cases, these unseen processes serve as the starting point for rational deliberation. Intuitions of justice, for example, tell us it's wrong to murder and steal. Reason expands these intuitions into libraries full of legal manuscripts.

Since the source of intuition is hidden, the ancients invented stories to explain how we obtained this implicit knowledge. Plato and other ancient philosophers believed that people were born with preexisting knowledge, and that learning is actually a form of remembering. It was thought that before birth, the soul descended through the seven planetary spheres, acquiring knowledge from each of the planetary gods as they passed. Intuition occurred when we accessed this divine knowledge.

The mystical meanings of numbers may lack the emotional intensity of fairy tales, but the ideas found in them continue to influence the way we think. These stories have lasted for so long, appear in so many cultures, and have attracted so much study because they are based on our shared archetypal blueprints. The meanings are not always the same in every culture, sometimes contradicting one another, as archetypes inevitably do, but the underlying themes are similar. As we look at the numbers, we'll include mystical traditions from different sources so we can understand the archetypes from more than one perspective.

NUMBERS AND NATURE

It's not surprising that numbers became containers for archetypal dramas. There's something inherently magical about them. They appear to hold nothing less than the secrets of the universe.

Typically, we advance our knowledge about the world through scientific experimentation. Researchers develop theories about how something works, then test those theories to see if they get the results they expect. A medical scientist might hypothesize that a drug developed for schizophrenia would work well for depression, then design an experimental study that would shed light on whether her hypothesis is correct.

But there's another way to learn about the world: numbers. Mathematicians can take formulas, manipulate them, and discover something beyond symbols on paper. Sitting in a room with nothing but pencil, paper, and their thoughts, they discover previously unknown truths about the material world. The Higgs boson, a recently discovered fundamental particle, was originally found using math. Only decades later, when the Large Hadron Collider was built, was an actual Higgs boson finally detected. The properties of this never-before-seen particle matched almost exactly what the math predicted.

There's a lively debate about why the material world so faithfully follows the rules of abstract numbers, but there's no consensus as to how or why it happens. Nobel Prize–winning physicist Eugene Wigner wrote a paper titled "The Unreasonable Effectiveness of Mathematics in the Natural Sciences."[1] In it, he says, "The miracle of the appropriateness of the language of mathematics for the formulation of the laws of physics is a wonderful gift which we neither understand nor deserve."

It's a gift that lies beyond reason. A hard-headed scientist who worked on the Manhattan Project during World War II, Wigner nonetheless chose to open his paper with a quotation from another Nobel Prize winner, mathematician and philosopher Bertrand Russell, that sets the stage for understanding numbers as something mystical:

> Mathematics, rightly viewed, possesses not only truth, but supreme beauty—beauty cold and austere, like that of sculpture, without appeal to any part of our weaker nature, without the gorgeous trappings of painting or music, yet sublimely pure, and capable of a stern perfection such as only the greatest art can show. The true spirit of delight, the exaltation, the sense of being more than Man, which is the touchstone of the highest excellence, is to be found in mathematics as surely as in poetry.

Math is the framework of modern science, the flower of human consciousness. But flowers need earth to grow, and numbers sink their roots deep into the unconscious. As archetypal stories, they contain a variety of meanings, both cosmic and personal, and provide us with a window into how the unconscious understands the universe and our place within it.

NUMBERS AS MYSTICAL ENTITIES

While there are, as noted, many traditions that assign mystical meaning to numbers, the main focus of this chapter will be the ideas derived from Pythagorean philosophy. Pythagoras was a Greek mathematician who believed numbers were real things, rather than just a convenient way of quantifying objects. Not only do we have two apples, two plates, and two people, but the number two also has its own existence. That existence isn't ink on paper. It's not the curved lines that depict the numeral "2." Rather, it is the idea of *twoness*, independent of objects to be counted or the numeral's written representation.

Designating *twoness* as a real thing may seem strange, but Pythagoras and other ancient philosophers believed the immaterial world of ideas was more real than the natural world of physical objects. The most famous example of this principle is seen in Plato's allegory of the cave. Plato described objects that appear to be real as nothing more than shadows projected on the wall of the cave, vague representations of the truly real, which are found in the realm of ideas.

Philosophers who believed that ideas were more real than physical objects based their theory, in part, on the observation that some things are perfect while other things are imperfect. Take a straight line. If you imagine a line in your mind, it's perfect. But if you try to draw one on a piece of paper, it will be imperfect. Even if you use a ruler, it won't be perfectly straight because of imperfections in the paper or just the bumpy nature of the molecules that make up the ink. The idea is perfect; every instance of the idea that gets translated into the material world is imperfect.

These philosophers also noticed that everything in the material world is constantly changing. The Greek philosopher Heraclitus pointed out that you can't step in the same river twice. The river isn't the same for even two seconds in a row. The same is true of an orange. It's constantly in the process of becoming a ripe orange, then a rotten orange, and eventually dust. Everything in the material world is in the process of *becoming* something else. For this reason, the material world was called "the world of becoming." On the other hand, abstract ideas, such as the idea of a line, are unchanging. The ancient philosophers contrasted unchanging abstractions with material things by calling abstractions "the world of being."

It's not surprising that they considered the imperfect, decomposing world of matter to be inferior to the perfect, eternal world of ideas. The Greek philosopher Nicomachus of Gerasa, in his "Introduction to Arithmetic," described it this way:

> [Numbers] are immaterial, eternal, without end, and it is their nature to persist ever the same and unchanging, abiding by their own essential being, and each one of them is called real in the proper sense. But [things in the natural world, which are] involved in birth and destruction, growth and diminution . . . are seen to vary continually, and while they are called real things . . . they are not actually real by their own nature; for they do not abide for even the shortest moment in the same condition but are always passing over in all sorts of changes.

Another Greek philosopher, Plotinus, writes, "Number exists before all objects which are described by number. The variety of sense objects merely recalls to the soul the notion of number." The word *recalls* is important. It derives from the belief that there was a time when the soul had a more direct connection to the idea of number. That's how it feels when we come into contact with the archetypes—as if we're encountering something we knew long ago.

Understanding numbers as disembodied entities—as having an existence independent of things that are counted—connects them to the magical traditions that tell stories about otherworldly spirits. Like the gods and goddesses of ancient mythology, numbers have distinctive characteristics and personalities. A mystical understanding of numbers is another doorway into the hidden chambers of the human psyche.

The most important numbers from a magical standpoint are the integers one through ten. They describe a cosmic progression that tells a story of birth, growth, death, and the transcendent return to unity. In many ancient traditions, everything begins with the One, the ouroboros that is the beginning and end of all things. Two is the catastrophic split that tears the One apart, and sets the stage for multiplicity—the One becomes many. Three is the first number of spirit. It gives intelligible form to unknowable prime matter and is also the spark that brings matter to life. Four is the number of matter, and the combination of three and four (form and matter) allow for the variety of objects that flow from the breaking apart of the One to be distinguished from one another and acquire identities. Five is the number of human consciousness, six the number of humanity's progress in the material world, and seven is the number of inner growth. Eight takes us into a mysterious world beyond reason, and nine is the final stage of the journey that leads us to ten, the holy number, worshipped by the Pythagoreans, which embraces both unity and multiplicity in a mystical transcendence.

ONE: THE ALL

One is the archetype of unity. One is the All. It's the thing that existed in the beginning, and it encompasses all creation. Everything flows from the One, and in the end, everything will return to the One. At the moment of the universe's birth, the moment of the Big Bang, the universe existed as a single point. There was nothing except that infinitesimal point, which contained everything.

The ouroboros is a symbol of the One. Before we awakened into consciousness, we were one with nature. In the end, when we die, we will return to that great mother. The light goes out and we rejoin the rest of nature in a single unity. One of the most famous depictions of the ouroboros is found in a manuscript written by Cleopatra the Alchemist. It carries the inscription *ἕν τὸ πᾶν*, Greek for "the All is One."

The One has a mysterious fascination for humans. It exercises a strong pull in all areas of thought, most obviously in religion and philosophy. You may have heard the joke about the Buddhist monk who asked a hot dog

Tenth-century alchemical ouroboros illustration, reproducing the work of Cleopatra the Alchemist, third century.
(Source: MS Marciana gr. Z. 299 via Wikimedia Commons.)

vendor to "make me one with everything." Natural scientists also feel the pull toward the One, toward a unified theory that will explain all scientific observations. In physics, the modestly named "Standard Model" reduces the near infinite number of things in the universe down to combinations of a small collection of elementary particles. Particle physicist Harry Cliff, who looks for signs of new physics at CERN's Large Hadron Collider, described the search this way: "There's a sort of obsession in physics, which is to try to unify things, to simplify complicated phenomena into one sort of single underlying phenomenon."[2] Some believe that string theory may take us to the one ultimate thing—vibrating strings—that gives rise to everything else.

TWO: AUDACITY

Two symbolizes the archetype of division. Two breaks the eternal perfection of one. For this reason, the Pythagoreans named it "audacity." Two also represents the emergence of consciousness out of the unconscious—a rebellion against the all-encompassing ouroboros. Like Lucifer, who said, "I will not serve," consciousness breaks free from unthinking obedience to nature and carves out its own kingdom, a kingdom characterized by thought, reflection, and choice. In mythology this process of awakening into consciousness, of freeing oneself from the cosmic mother of unconscious nature, is sometimes depicted by the slaying of the ouroboros, often in the form of a dragon.

We also see the awakening of consciousness depicted in the biblical story of Adam and Eve. Before the fall there was only good in the world. There was unity. Adam and Eve didn't have to struggle against evil aspects within themselves. They were instinctively, effortlessly good. Their will and the will of God were one.

They became aware of duality only after they ate the fruit from the tree of the knowledge of good and evil. Now they were able to see two sides rather than one. They had to make choices, because unity had been

In this illustration, the dragon, symbolizing the unconscious, is dark. The hero, representing consciousness, is illuminated by his golden armor, the halo around his head, and the sun. (Angelos Akotandos's *St. George on Horseback, Slaying the Dragon*, circa 1425.) *(Source: Wikimedia Commons.)*

fractured. Christian doctrine calls their act of eating the fruit "original sin," because it was an act of rebellion. But it's also been called *felix culpa*, Latin for "fortunate sin." It was called fortunate because it led to the miracle of the incarnation of Jesus. We can see it as fortunate in another way, as well. The story tells us that it gave humans the ability to choose whether they will be good or evil, an ability that can be viewed as conferring a nobility on the human race. Choosing to be good is much harder than involuntarily being good. By transitioning from unconscious perfection to conscious imperfection, the fall was, in a sense, a form of moral evolution.

The splitting of the psyche through conscious awakening has both positive and negative qualities. Although it raises humans to a higher level of existence via the acquisition of moral responsibility, duality is usually seen in a negative light, as we see in terms such as "two-faced," "duplicitous," and "double-dealing." In Edmund Spenser's epic poem *The Faerie Queene*, the good princess is named Una and the evil sorceress Duessa, names derived from the numbers one and two, respectively. Duessa's main characteristic is that she never appears to be what she really is. She's always practicing deceit; what she is and how she appears are two. In the case of Una, her appearance and her true being are the same.

As we saw in the chapter on the shadow, two represents not only conflict but also balance. It's a splitting into complementary polarities: masculine and feminine, night and day, sun and moon, matter and spirit. The archetypes unify opposites, which is part of the reason they can't be understood by the ego. When the opposites are separated, what was once incomprehensible chaos becomes intelligible order. The One is too much for the rational ego. Rationality requires duality.

The creation myth found in the book of Genesis can be viewed as the creation of order out of chaos by the separation of opposites. "Then God said, 'Let there be light'; and there was light. And God saw that the light was good; and God separated the light from the darkness . . . And God said, 'Let there be a dome in the midst of the waters, and let it separate the waters [of the earth] from the waters [of the sky] . . . And God said, 'Let there be lights in the dome of the sky to separate the day from the night.'"

Leaving the universal and looking at the role of duality in the individual, we see that division begins at birth. We leave the ouroboric paradise of the mother's womb and enter a cold world of comparative isolation. After the separation of birth, a second division takes place as the infant develops consciousness. The ego is split off from its instinctual origin. As we've seen, the process of growing up is a process of strengthening the ego. As they mature, children learn to think before they act. They are taught to suppress their instincts, to control their emotions, and to acquire the fearsome power of rational thought.

But in exchange for the power over our environment that we acquire by strengthening our ego, we lose the magic that gives life its charm. As Eugene Subbotsky observed in his son, we no longer feel that nature takes a personal interest in us. We're cut off from the roots of our origin, the deep, irrational unconscious. That loss creates a longing for return to the bliss of unity, whether through the oblivion of sleep or the stupor of intoxication. We desire to escape the conflict, complexity, and contradictions of the human condition, and so recover the simplicity of the One. But we cannot go back. We can only go forward.

THREE: LOVE

Three symbolizes the archetype of connection, spirit, and understanding. It's the first echo of one. Two split the unity of one into opposites. Three is the synthesis of those opposites—the reconnection. Because three binds together and creates a unity, it was called "love" by the Pythagoreans, the same word Empedocles used to describe chemical reactions in which two reagents came together to form something new, a third thing. When two people fall in love (a third entity that connects the two people), they may feel as if they have merged into a single whole. Love heals the split suffered by the mythological hermaphrodite. In a way, three justifies the terrible division of two. Restoring unity through love takes us to a higher level of unity than the blind, unconscious unity of the ouroboros.

In Western culture, the most well-known example of the unity inherent in three is the Christian Trinity, one God in three persons. Three represents wholeness in other ways, as well. We speak of beginning, middle, and end; birth, life, and death; morning, noon, and night. Three is also the beginning of understanding. It's when a pattern begins to emerge. As Ian Fleming writes in one of his James Bond thrillers, "Once is happenstance. Twice is coincidence. Three times is enemy action."

The geometric figure associated with three is the triangle. In the Greek alphabet, the triangle signifies the letter delta (Δ), which was associated with divinity. The Boeotians used this triangular letter to transform the Greek name for the king of the gods, "Zeus," into the word *Deus*, which became the generic Latin term for any of the gods and gives us the English word *deity*. Three is the number of the divine spirit.

Unlike a point, which is a unity, and a line, which is made from two points, the triangle is the first geometric figure that encloses an area and therefore can be called a shape. To put it another way, it's the first number to have a form. Recall that, for the alchemists, form was seen as a kind of spirit or soul. In this context, the word *soul* is used in a way that's different from how it's used today. In modern times, people who believe in the existence of souls grant them only to human beings and maybe some animals. But in ancient times the concept of soul was applied more broadly.

In its very broadest sense, everything has a soul because everything has a form that makes prime matter a specific thing. It is spirit, or we might call it *idea*, that infuses prime matter with identity and meaning.

A narrower concept of soul was used to distinguish living things from inanimate objects. The soul was the thing that gave inanimate matter its spark and brought it to life. In modern chemical equations, the Greek letter delta signifies heat or fire. It's the inciting factor that sets off a chemical reaction. In physics, delta signifies change of any type. It distinguishes a static environment from a dynamic one.

FOUR: MATTER

After idea (i.e., form) comes matter, which gives idea a physical existence. Four symbolizes the archetype of the material world. The number four is associated with matter because of the four elements of the ancients: earth, water, air, and fire. In geometric terms, a fourth point takes us out of the two dimensions of the triangle, up into the third dimension of the tetrahedron, a four-sided pyramid. By placing a point above the two-dimensional triangle, for the first time, we can make a solid. With four, we enter the world of physical existence.

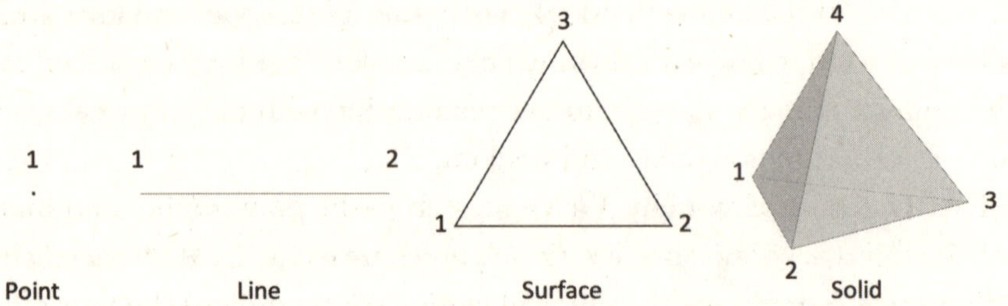

It takes four points to make a solid.

Four, being the number of matter, represents stability. It's associated with wholeness and justice ("a square deal"). There are four directions on the compass and four seasons in the year. The square, having four equal sides, is a geometric representation of stability. Mandalas, which symbolize unity and completeness, are based on squares or other four-sided shapes.

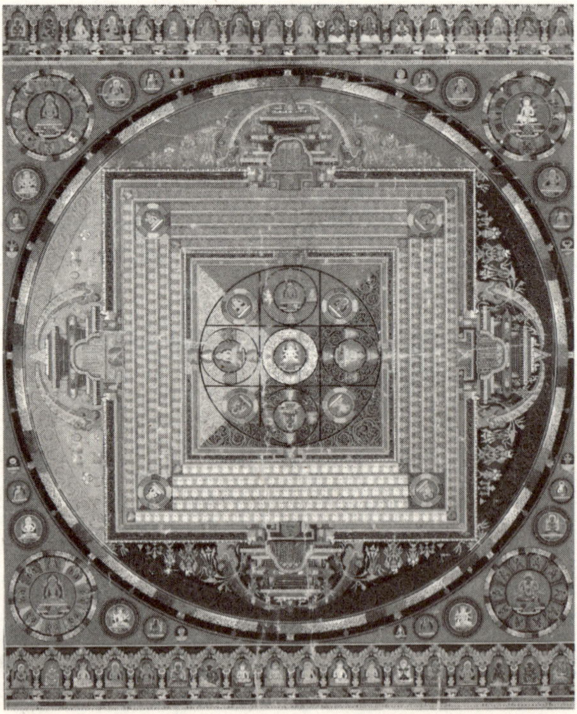

Anonymous Tibetan artist, *Mandala of Vajradhatu*, nineteenth century. *(Source: Wikimedia Commons.)*

In addition to the three heavenly virtues of faith, hope, and love, which involve our relationship with divinity, there are four cardinal virtues of classical antiquity that are oriented to our relationship with the temporal world: justice, fortitude, temperance, and wisdom.

Four is a place of stability. We've gone from the point to the solid shape, and from a geometric perspective, there's nowhere left to go. We have reached the four basic elements of the physical world. We encompass the four directions of travel and the four seasons of the year. We rest on the stability and completeness of the square. But we haven't yet reached our goal. What can move us forward? What can break us out of the solid stability of four?

FIVE: THE QUINTESSENCE

Five symbolizes the archetype of the human soul, the quintessence. *Quintessence* comes from the Latin *quinta essentia*, which means *fifth essence*. It's a divine substance that exists beyond the four physical elements. In English, *quintessence* is the finest, purest, and most concentrated essence of a thing.

In some ways, five is reminiscent of three, the number of spirit, but there is also a difference. Three is spirit in matter, the spark that brings inanimate matter to life. Five takes spirit to the next level. It's spirit that is aware of itself. Five is the human number. Whereas three has a strong echo of unity, five does not. Five is closer to the division of two. It combines conscious spirit with unconscious matter to create something strange and new—an animal that has been awakened into awareness.

Humans have five senses and five fingers on each hand. We also have four extremities plus a head, reminiscent of the four material elements plus the *quinta essentia*. Thus, the five-pointed star, the pentagram, is a symbol of humanity. When turned upside down it can be a symbol of the devil, representing fallen man, but it can also have the positive symbolism of the descent of spirit into matter, underscoring the strange union of conscious spirit and physical matter that is the human condition.

Another important shape associated with the number five is the "quincunx," a central element surrounded by four subsidiary elements. The quincunx highlights the spiritual element, the fifth essence, that gives soul to the four material elements. A cross is also a quincunx. Although we generally view a cross as a four-pointed shape, in medieval times it was seen as having

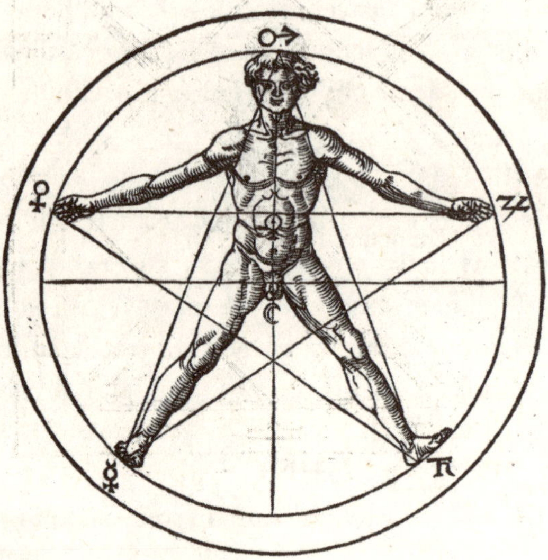

Heinrich Cornelius Agrippa, *Pentagram and the Human Body*, fifteenth century. *(Source: Wikimedia Commons.)*

five, because the intersection of the two arms was counted as a point. That point of intersection was the essence of the shape, often symbolizing a place of tension, conflict, and transformation or, alternatively, the cosmic center of the world.

SIX: THE EARTHLY PARADISE

Six is the number of material perfection. At the level of five, we noted that humans have both a mind and a body—spirit in matter. The mind judges between good and bad. The body acts on the physical world. The combination of judgment (mind) and the ability to act on that judgment (body) gives humankind the power to transform the world via productive work. Using our bodies, humans shape the material world until it matches what our minds judge to be good. The wilderness is transformed into a pleasing garden, and we attain six, the earthly paradise.

Six is what the Pythagoreans called a perfect number—one that is equal to the sum of its factors. In the case of six, $1 + 2 + 3 = 1 \times 2 \times 3$. Six is the smallest perfect number, and the only perfect number among the first ten

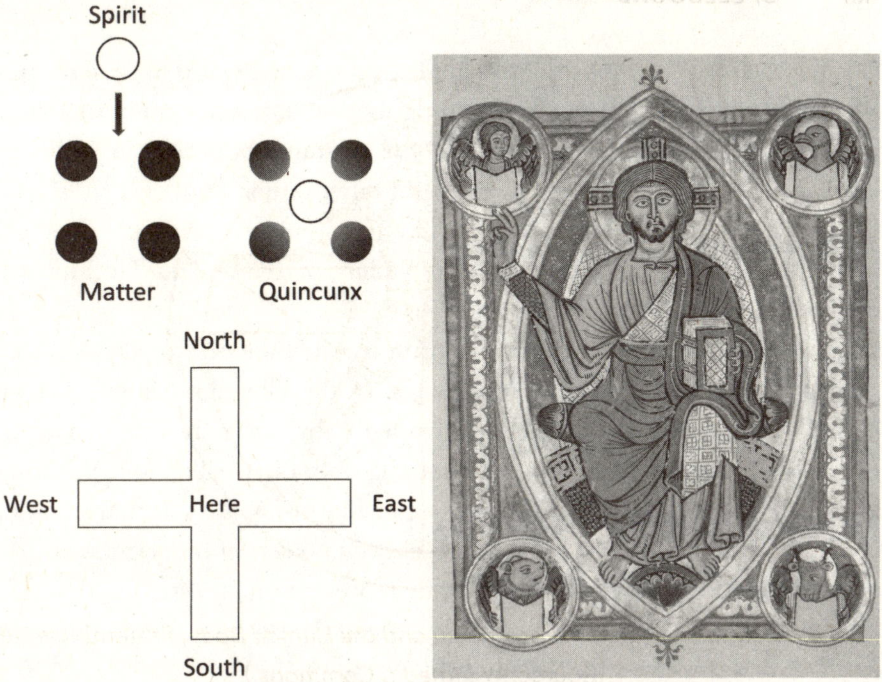

Examples of the quincunx and Christ surrounded by the symbols of the four evangelists.
(Source: Right, Evangelistar von Speyer, Codex Bruchsal 1 01v, c. 1220, via Wikimedia Commons.)

integers. Perfect numbers are rare. Despite the powerful computers available to modern mathematicians, only fifty have been discovered. After six, the next largest perfect number is twenty-eight. The fiftieth perfect number, the largest discovered so far, is $2^{74207280} \times (2^{74207281} - 1)$, which has 44,677,235 digits.

As a perfect number, six represents harmony and beauty and is therefore associated with Venus. It also represents the union of masculine and feminine. A triangle pointing down represents the feminine, and a triangle pointing up represents the masculine. When they are brought together, they form a six-pointed star, called a hexagram. In this aspect of six we see an echo of three—unity from duality—and therefore an echo of one. The hexagram became the symbol of alchemy because it represented the union of the macrocosm (the universe, symbolized by the upward-pointing triangle) with the microcosm (the human, represented by the downward-pointing triangle). As above, so below.

The book of Genesis tells us that God created the world in six days, and when it was over, "God saw every thing that he had made, and, behold, it was very good." St. Augustine writes, "Six is a number perfect in itself, and not because God created all things in six days; rather, the converse is true. God created all things in six days because the number is perfect."[3] Even this sainted church father couldn't resist the allure of six, making God himself subject to its fascination.

In the Jewish tradition, the hexagram is called the Star of David, referring to the ancient king of the Israelites. In the Christian tradition, King David is said to be the earthly precursor of the heavenly Christ. David's six-pointed star can be seen as representing both the perfection of his reign and the temporal nature of his kingdom, in contrast to the spiritual kingdom of Christ, who conspicuously rejected earthly power. In other contexts, the upward-pointing triangle more generally represents matter striving upwards, and the downward-pointing triangle represents spirit/form coming down to earth. The star is the union of the two.

Six also has negative associations. In the Book of Revelation, we're told that 666 is the sign of the beast. Many scholars have puzzled over what this sign means. There are many ways of interpreting it, and there's no agreement on which, if any, is correct. One possibility is to look at it through the lens of the earthly paradise. In the Christian tradition, the devil is the acknowledged prince of this world. He's said to draw his victims' attention away from the spiritual world by enticing them with material distractions.

Why the number six is repeated three times is also unclear. One explanation is that it's simply a diabolical mockery of the Holy Trinity. Another possible explanation is that it represents the perfection of physical enjoyment multiplied to sickening excess.

We see the negative side of the earthly paradise in Aldous Huxley's dystopian novel *Brave New World*. Science, technology, and social engineering have produced a seemingly perfect society of happily indoctrinated citizens. Everyone has all their needs satisfied, but they are slaves of the state. Huxley was less worried about the dangers of tyranny by force and more worried about the dangers of having our nobler desires washed away by trivial pleasures.

Once again, we've reached a place of stasis, as it's easy to get stuck at six. What could induce anyone to move beyond material perfection, the earthly

In this Jain Yantra, a mystical diagram used in worship, we see the four-sided square, representing wholeness; the quincunx, in which the god sitting at the center represents the quintessence inspiring the material world; and the hexagram, the perfect union of opposites. (*Vardhamana Vidya Yantra Pata*, fifteenth century, artist unknown.) *(Source: HereNow4U.)*

paradise where all of one's needs are met, and all physical delights are available? How does one escape the prison that this safe, enclosed six-sided cube can become?

SEVEN: TRANSFORMATION

Seven is the magic number, the archetypal number of transformation. Three represents the simplest form of soul, which gives life to inanimate matter. Five is the conscious soul that is aware of itself. Seven is the soul that is aware of something higher than itself.

Seven invites us to look up and see beyond the human world. When the ancients looked up at the night sky, they saw the seven planets embedded within their heavenly spheres. The hierarchy of the planets led to the expression "seventh heaven," meaning the highest state of bliss. In medieval times, there were thought to be seven steps to perfection, corresponding to a progression through the seven heavenly spheres. The number seven orients us to transformation and growth. When we look up, we see something higher than ourselves, an ideal we fall short of, which creates in us the desire to become better than we are. As the nineteenth-century mystic George MacDonald writes, "Instincts of betterment tell me I must rise above my present self."[4]

God created the world in six days, then set aside the seventh as a day of rest, a time to take one's mind off mundane work and contemplate higher things. The Roman Catholic Church recognizes seven sacraments, which represent turning points in a Christian's life. The three heavenly virtues plus the four cardinal virtues give us a total of seven paths of development. However, spiritual transformation is not always in the direction of progress. There are also seven deadly sins (pride, greed, lust, envy, gluttony, wrath, and sloth)—virtues such as self-confidence, aspiration, and innocent desire taken to pernicious extremes.

The idea of something higher than ourselves is what breaks us out of the earthly paradise of six. Hedonistic pleasures are all well and good, but they can't satisfy the human longing for wholeness. In *The Odyssey*, the story of the voyage of Odysseus as he makes his way home from the Trojan War, he and his men are blown off course and arrive at the land of the lotus eaters. The lotus is described as so delicious that anyone who tastes it enters into a state of bliss and no longer cares about returning to their home. Odysseus refuses to succumb to these pleasures and forces his men to pursue their journey. Later, he comes to the island of Circe, a beautiful goddess and enchantress who falls in love with him. Although he spends a year feasting and drinking with her on her island, eventually he is overcome by the desire to return home. Even the promise of life with a goddess is not enough to sway him from his journey toward his ultimate goal.

Whereas the perfection of six is static, the dynamism of seven drives us down the path toward wholeness. It's a path that's often painful. In order to become someone new, we must allow the old self to die, like the phoenix bursting into flames. Think of the child who must become an independent adult. His old identity must die before the new one can be born. The state of dependence, of being taken care of, must undergo the often painful transformation into independence and taking care of others. Even trifling transformations, such as giving up eating ice cream before bed, can feel like the death of some small part of ourselves—the part that habitually anticipated the pleasurable indulgence.

It's uncomfortable to be an imperfect combination of matter and spirit, a chimera of the animal and the divine. The story of Odysseus reflects a universal feeling that we need to find our way home, that we're not yet at the place where we belong. As Jung writes, "The heart glows and a secret unrest gnaws at the roots of our being."[5] We instinctively know that our journey is one of transformation. We are not yet where we need to be, and we're not yet the person we were meant to become. Seven symbolizes the difficult path that points the way toward wholeness and perfection.

EIGHT: TRANSCENDENCE

The One cannot be conceived by the human mind. Two and three are also difficult; they represent concepts that are too large. Four, five, six, and seven are more comprehensible. But when we reach eight, we once again go beyond the ordinary level of human understanding. Eight is transcendence.

Eight is associated with return, but a return that takes us to a higher level. There are seven days of the week. The eighth day is a return to the first day of the next week. The Western musical scale contains seven notes. The eighth is the octave, which returns us to the first note, but at double the frequency. In medieval times, the baptismal font was an octagon, symbolizing rebirth on a higher level. Christ's resurrection took place on the day following the Sabbath—the eighth day of the old week or the first day of the new one. The feast of Easter is celebrated for eight days and is called the Octave of Easter.

The higher level of eight is different from the mundane world we're used to. We've encountered the astrological myth that when a soul descends to Earth it passes through the seven spheres of the planets and acquires characteristics from each of the planets. Macrobius, a fifth-century Roman scholar, writes that souls acquire contemplative power from Saturn, executive power from Jupiter, valor from Mars, and so on.[6] According to the alchemical *Corpus*

Hermeticum, after death, the soul retraces this journey in reverse, ascending back through the seven spheres and returning to each planet its virtues and vices. At last, stripped of all temporal attributes, it travels beyond the seventh heaven to the sphere of the fixed stars. The soul has reached a place of peace. In contrast to the constantly moving spheres of the seven wandering planets, the eighth sphere is eternal and changeless. The Chinese philosopher Lao Tzu refers to the importance of stripping away the nonessential on the path to personal growth when he writes, "To attain knowledge, add things every day. To attain wisdom, remove things every day."

Modern society is best characterized by the number six. We're focused on solving the problems of the material world, particularly the problem of resource scarcity. Using technology, we're moving toward engineering the earthly paradise in which every material need will be met. As individuals, most people live their lives at the level of seven. We look up and become aware of something more perfect than ourselves, whether a deity or an ideal, and we aspire to make ourselves better. But achieving the transcendence of eight requires a lifetime of growth and experience.

The sayings of the enlightened tend to be characterized by the resolution of paradoxes and the uniting of opposites, along with nonjudgmental acceptance of the world. Here is a Zen verse that illustrates the idea:

In spring colors, there is neither high nor low,
The flowering branches are, by nature, some long, some short.
An ordinary person knows it and becomes a sage;
A sage understands it and becomes an ordinary person.

With eight, we have now attained the greatest stability of all: the peace of the sphere of the fixed stars. What force could be powerful enough to dislodge us and move us along the final stage of our journey?

NINE: THE DARKNESS BEFORE THE DAWN

What lies beyond the sphere of the fixed stars? What is beyond the outermost limits of what we can see when we look up into the sky? Beyond the mythological eighth sphere, all is darkness. Nine is the end. It is the last in the series of primary integers. Nine is death.

There are nine levels of Mictlan, the underworld of Aztec mythology. The river Styx circles Hades, the underworld of Greek mythology, nine times. Christ died at the ninth hour of the day, and the Norse god Odin hung for nine days from Yggdrasil, the tree of life, on which he sacrificed himself.

The seven-eight-nine series is a common mythological motif, typically involving the archetype of the wise old man. The original Star Wars trilogy uses the seven-eight-nine series to tell the stories of both Obi-Wan Kenobi and Yoda. People in stories who have reached transcendence have nowhere left to go in terms of character development. The only thing left to do is die.

TEN: TRANSCENDENT UNITY

The numbers one, two, and three are the numbers of creation. Four, five, and six are the numbers of what has been created. Seven, eight, and nine are the numbers of the return to the One in a new form. One is the creative cause of the cosmos; ten is its full development and completion. It is one from many, *e pluribus unum*.

Three healed the division of two with the addition of a third thing, love, which recombines. Ten represents a similar process on a cosmic scale. It is the universal love that unites all of creation. It's also a mystical paradox because, within its unity, the individuality of each member realizes its highest potential. All is One, while each is unique.

The importance of the number ten to the Pythagoreans can be seen in the mystical tetractys, a figure in the shape of an equilateral triangle consisting of ten points arranged in four rows. The Pythagoreans believed it held the secrets of the cosmos. Iamblichus, a first-century Neoplatonist philosopher who wrote a biography of Pythagoras, described the tetractys as "both a mathematical idea and a metaphysical symbol that embraces within itself—in seedlike form—the principles of the natural world, the harmony of the cosmos, the ascent to the divine, and the mysteries of the divine realm."[7]

It may not look like much, but the figure was so important, the Pythagoreans would swear binding oaths on it and even pray to it:

> Bless us, divine number, thou who generated gods and men! O holy, holy Tetractys, thou that containest the root and source of the

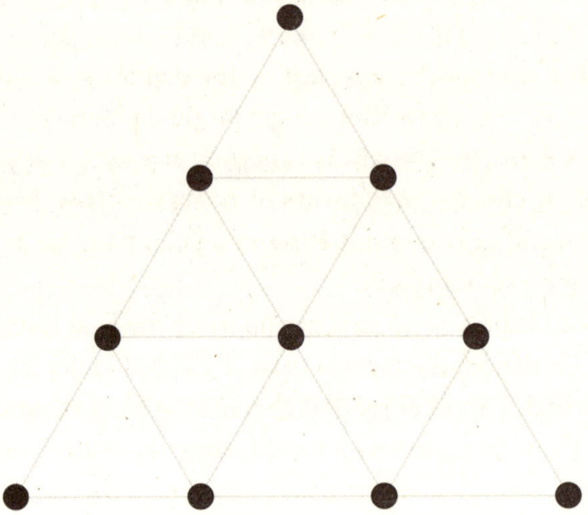

The tetractys of Pythagoras.

> eternally flowing creation! For the divine number begins with the profound, pure unity until it comes to the holy four; then it begets the mother of all, the all-comprising, all-bounding, the first-born, the never-swerving, the never-tiring holy ten, the keyholder of all.[8]

Like the mandala, the tetractys is based on the principle of four, the number of wholeness. There are four rows, and the figure stands on a base of four points. The three blank rows between the points were also important, because they represented ratios, specifically 1:2, 2:3, and 3:4. These ratios define the most important intervals in music. Two strings vibrating with a ratio of 1:2 produce an octave, 2:3 the perfect fifth, and 3:4 the perfect fourth. These intervals are naturally pleasing to the human ear. To them, Pythagoras added four additional intervals to match the number of planets in the heavens, and this led to the diatonic scale on which most of today's music is based. We may not be able to hear the legendary music of the spheres, but the planets make their influence felt whenever this seven-note musical scale is used.

From a cosmic perspective, ten symbolizes the unification of all things. For the individual, ten represents the harmonious integration of all aspects of the psyche, and thereby the creation of a whole, unique, fully individuated self.

EXCELSIOR

There are times when we're tempted by the comforts of stability, when it's easier to rest than to struggle. But struggle is the human condition. It doesn't matter where we are, there is always another step to be taken, and we won't find happiness by clinging to old ways of being that have become worn out. We live in the world of becoming. If we don't climb up, we're destined to fall down. There is no standing still.

In the next chapter, we'll look at one more magical tradition that tells a story of transformation: the tarot, a deck of cards that depicts the archetypal stages of growth. It's another map that charts the hidden paths of the unconscious and helps us find our way forward along the path to becoming whole.

THE TAROT

The true Tarot is symbolism; it speaks no other language and offers no other signs.

A. E. WAITE

A GAME OF CARDS

The tarot was created in the fifteenth century in northern Italy when a fifth suit of trumps was added to the usual four-suit deck. The new deck was originally called *carte da trionfi*, cards with trumps. It was later changed to *Tarrocchi*, a name that artist and tarot historian Robert Place speculates came from the name of the river that drove the paper mills where the cards may have been manufactured.[1] Originally, the new deck was used to play a game similar to modern bridge. It wasn't until the eighteenth century that occultists popularized its use in divination.

The tarot was illustrated with ancient mystical symbols that had been rediscovered during the Renaissance, when the cards were created. None of the symbols are unique to the tarot; they can be found in many other works of popular culture from that era. Artists used these symbols to give their work both body and soul. The body was the subject being portrayed; the soul was the deeper meaning the subject symbolized.

Unlike alchemy, which was intentionally made obscure to hide its secrets from the uninitiated, the tarot was designed to be accessible by all. During the Renaissance, people made no distinction between high art and popular art, and so a simple game of cards was illustrated with symbols of ancient philosophical mysticism. The use of these symbols was so widespread that people were as familiar with them as we are with our own cultural icons, like Santa Claus. As cultural journalist Josh Jones writes, "Literate or not, Medieval Europeans spoke a language of symbols that stood in for whole folk traditions and theologies."[2]

There was a great deal of inventiveness and competition among artists during the Renaissance, and consequently, many different tarot decks were produced containing a variety of symbols. Like fairy tales, the tarot became a collective work. With each new deck, some images were retained and modified, others abandoned. In the sixteenth century, alchemical and Pythagorean symbols were incorporated into the deck by French artists, leading to the creation of the Tarot of Marseilles, which most modern decks are based on.

The most popular deck today, the one we'll focus on, is the Waite-Smith, which was first published in 1910. It's sometimes called the Rider-Waite deck, referring to its publisher, the Rider Company. The cards in the Waite-Smith deck were created by illustrator Pamela Colman Smith, based on instructions sent to her by occultist Arthur Edward Waite. The cards are derived from the Marseilles deck, with some modifications to the trump cards, and more extensive illustrations given to the cards in the four familiar suits. Waite called the suits the "minor arcana" ("arcana" means secret or mystery) and the trumps the "major arcana" or "keys."

The cards are strikingly beautiful. Artist Sharmistha Ray describes them as "archetypal subjects that each become a portal to an invisible realm of signs and symbols, believed to be channeled through processes of divination."[3] A Jungian might restate the last part as "channeled through eruptions of unconscious archetypes."

Éliphas Lévi, a French occultist whose writings influenced Waite, writes, "An imprisoned person with no other book than the Tarot, if he knew how to use it, could in a few years acquire universal knowledge, and would be able to speak on all subjects with unequaled learning and inexhaustible eloquence."[4] You may recall from the previous chapter the ancient belief that learning was believed to be a process of remembering the knowledge that was given to us prior to our birth as we descended through the spheres of the seven planets. Lévi believed the tarot could unlock that knowledge buried inside us.

A psychological approach might credit the symbolism of the tarot with giving us the ability to make conscious some aspects of our unconscious. Just as a photographer needs to be familiar with different kinds of lenses in order to understand how light becomes a photographic image, we need to be familiar with the components of our psychological structure—the archetypes—to understand how our interactions with the world make us who we are. Archetypes are activated during numinous experiences and in dreams, but, under ordinary circumstances, the best way to encounter them is through symbols like the ones collected in the tarot, physical representations that can be perceived by the senses. These symbols personify the archetypal forces to make them more accessible to conscious understanding, and help us recognize them in our own lives so we can manage them more effectively. That's not as dramatic as containing all the world's knowledge, but as the Oracle of Delphi reminds us, knowledge of oneself is the most essential kind.

Modern people are unfamiliar with most of the tarot's symbols, but we can correct this deficiency by analyzing the cards—by taking them apart and identifying each symbol. However, it's important to keep in mind that an analysis of this kind won't tell us what the cards mean. Because of their symbolic nature, the meaning of the cards will be unique for each individual and will change with the changing circumstances of his psychological state.

Once the symbols have become familiar, the way to understand the meaning of a card is to allow the symbols to unlock images and ideas in the imagination. The message needs to emerge from the unconscious, which means that it occurs outside conscious control. The meaning may leap out from the cards, it may take days to develop, or it may seem that nothing at all comes in response to the cards. When dealing with the unconscious, the ego has to trust. The long, winding path that lies between the symbols and a consciously experienced impulse, idea, or emotion can make the connection between the two hard to see.

THE KEYS

In a tarot deck, there are twenty-two major arcana, or keys, in addition to the four suits found in a traditional deck of cards. There are many ways to interpret these cards, using not only classical Western philosophical ideas but also the Kabbalah and Eastern mysticism. We'll look at the symbols from a Jungian perspective, as a metaphorical representation of the soul's journey toward individuation. Robert Place suggests that each of the keys represents a way station that must be mastered before the soul can move on to the next level. Closely examining the first five keys, as we'll do here, will give us an introduction to the ways in which the symbols help us become more familiar with the unconscious archetypes. (Descriptions of the remaining seventeen keys can be found in the appendix.)

0: THE FOOL

The Fool is the first of the keys and stands apart from the others. It's given the Arabic numeral zero, in contrast to the other keys, which are all given Roman numerals. In the traditional card game that was played with the deck, the Fool was a wild card. In some ways, the Fool isn't part of the progression toward wholeness. Since its number is zero, we haven't yet begun.

The card depicts a youth full of beauty and grace. The expression on his face is serene. He carries a flower in his hand, a symbol of fertility. The sleeves of his tunic are like flames. His companion is a dog, symbolizing his easy alliance with his animal nature. The sun, the symbol of consciousness, shines brightly in the sky, but the youth's eyes are closed. He is Adam before the fall, effortlessly perfect, his eyes not yet opened by the fruit of the tree of knowledge. He carries a bindle on his shoulder, showing that he's going on a journey, and we can see where the journey will take him—over the edge of a cliff. The Fool is about to enter the world of becoming, where he will lose his natural perfection. He will have to fight to earn it back, but in the end, he will regain it at a higher level. The fall will be his first step toward realizing the potential of his soul. Without the pain and suffering of the journey, he will remain unconscious of his gifts.

Why is this noble young man called a fool? The reason is that he represents the beginning of a new endeavor. To learn something new, we must begin with both an open mind and the courage to appear foolish. If we come in as an expert, we'll have so many preconceived ideas, we won't see anything that lies outside our existing model of how things ought to work. This is the advantage of the amateur, the ability to see things from a fresh perspective. Charles Darwin was an amateur scientist who never held an official position in his field of study. So was Gregor Mendel, an Augustinian friar who was the founder of modern genetics. To grow beyond our current limitations, we must adopt the mindset of the amateur, the attitude of the fool who knows nothing. It will inevitably lead to feelings of insecurity and repeated episodes of failure, but those who play it safe will undergo the punishment of petrification. To make progress, we must take risks.

I: THE MAGICIAN

THE MAGICIAN

Magic occurs when the natural world is invaded and occupied by entities from the spirit world. Although materialists like Francis Crick tell us we're nothing but a pack of neurons, from a subjective point of view the human mind feels like it's one of those spiritual entities. It's a strange sort of spirit, though, because it's attached to a body, connected to it by a bond of magic. Despite having no fermions or bosons, the immaterial mind magically influences the physical body, and through the body, the world.

The Magician connects spirit and matter. The intentions of the ego become decisions that set the body in motion, decisions that may have important consequences. The Fool has entered the world of becoming and must now make choices about what to do. He must exercise his will. The Magician points upward with his right hand and downward with his left, portraying the alchemical maxim "as above, so below." The Magician has a body and a soul, and he knows he must use both, ultimately unifying them in a state of harmony.

Lying on a table in front of him are the instruments of magic: the wand, cup, sword, and pentacle. These are the mystical tools he must learn to use to achieve successful growth. The Magician has not yet picked them up to begin his work, but he will. Unlike the Fool, who exists effortlessly, the Magician must engage with the world. He must labor along the path of transformation.

Above his head is a lemniscate, the sign of infinity. It reflects reason's ability to comprehend abstract ideas—eternal truths from the world of being. Around his waist is the ouroboros, which links him to the natural world. Even though consciousness lifts him above ouroboric darkness, it's still an essential part of his humanity and must be retained.

The magical instruments on the table represent the four suits of the minor arcana: wands, cups, swords, and pentacles, corresponding to the clubs, hearts, spades, and diamonds of a modern deck of cards. They're associated with the four elements of antiquity. The wand is drawn as a staff, such as a wizard might carry. It's associated with fire. As we proceed along our journey, fire provides inspiration, sparking new ideas. It's the agent of creation. Each new stage along the journey will require new ways of understanding ourselves and the world. Our experience from previous stages will help, but there will be unfamiliar challenges that require novel solutions. Analysis can't unravel these problems. We need the help of ancient instinct to supply new adaptations through inspiration and intuition.

The cup represents water. A cup is a receptacle. It receives and holds the inspiration of fire so that it is not uselessly dissipated, an idea that is forgotten. Some writers carry notebooks around with them because they never know when inspiration will strike. If they don't catch it, it will be lost. The notebook is their magic cup. Sometimes a new idea, if it's good enough, stimulates emotion, and that helps us remember. Emotion is also a kind of cup, and as we saw in the chapter on alchemy, it is also associated with water. Like the magic cup that captures fire, emotions trigger the brain to encode fleeting experiences into long-term memory. Up to a point, the stronger the emotion, the stronger the memory. (Traumatic experiences can metaphorically break the cup and cause amnesia.)

The sword is associated with air, which is in turn associated with the intellect. It cuts things into pieces so they can be analyzed. The blade also cuts away falsehood, separating truth from error. As a symbol of the intellect, the sword also represents consciousness. It's notable that the symbol takes the form of a weapon, reminding us that the ego must constantly fight for its existence, resisting the power of the ouroboros that would absorb the ego back into itself.

Pentacles represent earth. They symbolize the transformation of mental activity into something that impacts the physical world. Will is exercised, a decision is carried out, and an idea becomes a reality. The Tarot of Marseille uses coins as the fourth suit, symbolizing the wealth that buys material goods. The Waite-Smith deck retains the round shape and golden color of the coin, but imprints it with a pentagram, the five-pointed star that symbolizes humanity. Human spirit stamps its impression on the material world.

II: THE HIGH PRIESTESS

The High Priestess is the ego's connection to the unconscious. She's the gateway to the hidden springs of life that stream upward from the darker regions of the mind. At her feet is a crescent moon, symbolizing night and water, prime matter. Her gown tumbles at her feet like a waterfall. Behind her are pomegranates, a symbol of fertility because, when opened, they burst with seeds. She sits on a simple cube of stone, another symbol of prime matter.

Because the High Priestess connects the two components of the psyche, we see not only symbols of the unconscious, but also symbols of unification. There are two pillars marked with the initials B and J, representing Boaz and Jachin, the two pillars that stood on the porch of Solomon's Temple. One is light, the other dark, and the High Priestess sits between them as mediator. She wears the crown of Isis, which combines the sun with the moon. The solar cross on her chest complements the moon at her feet and represents spirit in matter. The vertical line of the cross is spirit coming down from above. The horizontal line is matter, penetrated by spirit. The intersection, the quincunx, is the sacred place where they meet.

To reach our goal, the attainment of full individuality, we're going to need all our abilities. The first step is to establish a line of communication with the hidden regions of the mind. The High Priestess fulfills this role.

III: THE EMPRESS

Once we have established a connection to the unconscious, the source of life, interest, and enthusiasm, we turn our attention to the outside world, where *eros* finds objects to desire. The Empress symbolizes the archetype of bountiful nature. She produces the things that sustain us and give us pleasure. She is Venus in her role as the goddess of fertility.

It's tempting to feel contempt for the lower steps along our journey through the keys. But that would be a mistake. All the stages of the journey are essential. Desire may be viewed as a primitive emotion, reflecting our animal nature, but without desire, we can't make intellectual sense of the world, because assigning value to a thing is an essential part of understanding it.

One of the symptoms of clinical depression is anhedonia, the inability to experience pleasure. A good meal, a good book, lunch with a friend—for the person suffering from anhedonia, all of these things are devoid of pleasure, and this robs the sufferer's life of both joy and meaning. We get up in the morning and begin our work to attain the things we desire, whether a happy family, a paycheck, or the satisfaction of helping others. If there's nothing you want, nothing you desire because nothing gives you pleasure, how do you decide what to do? How do you assign value—meaning—to the things around you? When nothing gratifies, everything is the same. The world is flat, gray, and dull. We need our unconscious to produce desire and give things meaning, which allows us to understand the role things play in our lives. The flames of passion produce not only heat but also light.

The Empress card shows archetypal desire sitting in luxury, surrounded by the bounty of nature. Her head is crowned with twelve stars, representing the months of the year. Her scepter is ornamented with a globe of the world, her dominion. Her role as goddess of fertility is emphasized by the lush, green vegetation behind her and the copper-colored grain, ripe for the harvest, growing at her feet. A heart inscribed with the sign of Venus rests against her throne, representing *eros*.

IV: THE EMPEROR

Desire leads to action. The High Priestess gave us access to our instinctual drives, and the Empress provided those drives with objects to love and desire. The next step is initiative—setting out to obtain the things we want. The Emperor is clad in armor, ready to fight, to take on the obstacles that stand in the way of his goals. He is the complement of the Empress. While she symbolizes that which attracts desire, she herself is unmoved. The Emperor, by contrast, is desire in motion. He actively pursues the things he wants.

The landscape behind the Emperor is rocky and barren, in contrast to the lush growth behind and around the Empress. Whereas she is soft sensuality, the Emperor is hard discipline. The corners of his throne are decorated with rams' heads, symbols of earth. In Christian lore, the ram's head has become a symbol of the Devil, because he's considered the prince of the fallen world. But this is a stolen sovereignty, and so the ram is not fundamentally connected with the Devil. It began as a symbol of the masculine aspects of nature, complementing the feminine symbols of the Empress.

There's joy in achievement, the realization that we can effectively impose our will on the world to obtain the things we desire. People often first experience this ability in their twenties, when they have completed their education and are ready to take on the world. The Shakespearean character Pistol, faced with a difficult mission and deprived of assistance, remarks, "Why then the world's mine oyster, / Which I with sword will open." He won't allow obstacles to stand in his way. He will crack open the world's hard shell to win the prize it holds. He acts under the influence of the Emperor.

USING THE TAROT

The tarot deck's twenty-two keys—the five we've reviewed here, and the seventeen in the appendix—contain symbols and personifications of instincts and archetypal events that are common to all humanity. They represent the wisdom of the ages, a collection of images that have inspired men and women from millennia ago to the present. How can we make use of this wisdom to enrich our lives and help us along the path of individuation and growth?

As we live our day-to-day lives, archetypes are activated within us. They influence the way we understand our experiences and how we feel about them. This influence usually occurs outside conscious awareness. For instance, we might feel an immediate connection with someone we just met and have no idea why. Or, more likely, our ego will invent a rational reason to account for the irrational feeling of connection. We might say, *I like that person because she's smart or funny*. But we've met dozens of smart, funny people before without feeling that indescribable connection. In most cases, something else is going on.

The most obvious clue that archetypal activation is occurring is a non-ordinary state of consciousness, for example when our emotional reactions to other people seem out of proportion with what they're actually saying or doing. Under the influence of archetypal projection, we often feel a little off balance and may notice we're having more trouble expressing ourselves than usual. Our words are clumsy, our thoughts are jumbled together, and we may blurt out things that are uncharacteristic of us. *Logos* is eclipsed by *eros*.

If we pay attention to these things, we have an opportunity to recognize one of the archetypes: a god, goddess, spirit, or demon who has emerged from the darkness and cast a spell over us. We might get in the habit of asking, "Who else is in the room?" Familiarity with these archetypes lets us recognize them more readily, and provides an opportunity to gain a deeper insight into our own behavior. The more we know about the archetypes, the better we'll understand the spells they cast. We'll also have a better understanding of the problems that confront us when we encounter a particular archetype, as well as the risks and possibilities we may encounter. Any time an archetype is activated, there is both danger and opportunity. It's good to prepare yourself.

The tarot is an encyclopedia of these archetypes. It allows us to better understand the archetypes' dynamics, so that, rather than being blindly buffeted by the tempest of unconscious agents within us, we can join forces with these hidden partners to accomplish the goal of inner growth. The tarot acts as a kind of guidebook, telling us where we are, who we're dealing with, and what to expect.

A hiker walking through the woods is more likely to reach his goal safely if he understands the ways of forest. The experienced hiker knows that camping among a dense thicket of trees will provide protection from the wind. He knows that if he builds his fire in front of a large rock it will radiate heat back into his tent that would otherwise be lost. And if he's in need of water, he knows to look for animal tracks, because they often lead to a stream or a lake. The more he knows about his surroundings, the more successful he'll be. It's the same with life. As we grow, we face unfamiliar challenges. If we want to navigate those challenges successfully, it's helpful to have a map that shows us the landmarks, the dangers, and the opportunities we'll encounter.

For some people, the tarot is more than a collection of mystical symbols that reflect the inner workings of the psyche. They believe a tarot reading, in which cards are selected at random and arranged in specific patterns, can answer questions, provide guidance, and even predict the future. Ordinarily, tarot cards are read by expert readers. These readers tend to be more intuitive than most, and may unconsciously notice a querent's reaction to the cards, which then guides their interpretations. But there may be some advantages to reading one's own cards.

A querent might look through them, one at a time, until she notices she's lingering over one. Maybe there's something interesting about it, or even fascinating, a sign that projection is taking place. That's the one that will have the most to say at that particular moment. Alternatively, one of the cards may produce a feeling of discomfort or anxiety. So long as the feeling can be tolerated, that's also a card worth spending time with. The richest treasures are guarded by the most frightening beasts. The next step is to contemplate the card, examining the details that appear most interesting or anxiety provoking, staying on the lookout for spontaneous thoughts or feelings or other signs of unconscious activity.

Imagine someone feeling tired, worn out, and cut off from the flow of life. He picks up the cards, scatters them over a table, and pushes them around with his hands. After a while he notices that the Nine of Wands (one of the minor arcana) keeps capturing his attention, and he decides to spend some time with that card.

Not knowing his past or personality, we can't be sure how the card affects him, but maybe the wands make him think of fire, perhaps the absence of it in his life; there's no movement, no new ideas sparked by inspiration. Its number is nine, the number of darkness before the dawning of the mystical light of ten.

These associations give him hope. Maybe he's not at a dead end. Maybe his psyche is just preparing for some kind of rebirth. He's a realist; he knows it can take years for the light to dawn. But knowing he's in the realm of the Nine of Wands helps him understand the darkness that surrounds him. The world may seem desolate, but that's because desolation prepares the way for change. Death must precede new birth. By orienting himself to the bigger picture, he gives the darkness meaning. He knows it's there for a reason, and that makes it more tolerable, maybe even productive. By adopting the perspective of the unconscious, channeled through the symbols of the card, his negative conscious attitude becomes more balanced.

Symbols express what thoughts cannot think. We cannot know what a tarot card—or an alchemical text, or a fairy tale—means. We can only open ourselves to its influence, and if we're lucky, that influence will reveal something valuable—something that connects us to our unconscious partner and carries us another step along the path of individuation, the realization of the unity of the self. The path is a rocky one, filled with false turns and dangerous pitfalls, but it leads to the heights of human existence. It's this path, and its end goal, that we turn to next: the path of transcendence.

PART III

TRANSCENDENCE

BECOMING TRANSCENDENT

Only to the extent that someone is living out this self transcendence of human existence, is he truly human or does he become his true self. He becomes so, not by concerning himself with his self's actualization, but by forgetting himself and giving himself, overlooking himself and focusing outward.

VIKTOR FRANKL

TRANSCENDENCE DEFINED

The ultimate goal of the mystical journey is *transcendence*, a word that means "to go beyond." In the mystical tradition, transcendence refers to rising above the apparent duality of opposites.

Symbols of transcendence are common in magical traditions. In the progression of the integers, we find individual transcendence in the number eight as well as cosmic transcendence in the number ten. In fairy tales, transcendence is often symbolized by a royal marriage or in the paradoxical unity of opposites found in magical creatures. Ugly toads transform into beautiful lovers, while filthy beggars and common animals perform feats of miraculous power to help those who trust them. Transcendence in alchemy is symbolized by the joining of the higher with the lower. Base metals contain the latent capacity to be transformed into noble gold, and correspondences unite the personal with the cosmic.

In Jungian psychology, transcendence is seen in the unity of opposites contained within the archetypes, the joining of darkness with light and the spiritual with the chthonic. It's also seen within the potential of the self to transcend the duality of the ego and unconscious—the process of individuation, which Jung saw as the key to personal growth.

Most research into human development has focused on children. An understanding of the expected stages of growth allows parents, teachers, and healthcare professionals to identify children who are struggling and find ways to help them move forward along the road toward becoming successful adults.

But personal growth doesn't end when we reach adulthood. It continues throughout our lives, albeit at a slower pace and in more varied ways. The early stages of development, such as acquiring the ability to sit, walk, and use language, proceed along a predictable path for most children. But adult development—the process of individuation, which makes a person increasingly different from everyone else—doesn't follow a well-defined course. That raises the question: What indications of growth, exactly, are we looking for?

Most people agree that successful development as an adult includes the ability to establish and maintain long-term relationships, and to become independent—to work and care for oneself. A more ambitious goal is finding ways to fit in with the expectations of society while also cultivating individuality. But some researchers look even higher than this, to the realization of transcendence.

Experimental psychologists who study human growth use the term *self-transcendence* to describe the ability to move beyond a primary focus on oneself and adopt a broader perspective on the world. Self-transcendent people look outward, away from themselves, prioritizing friends and family, humanity, other species, and the universe as a whole. The *self* in *self-transcendence* refers to the conscious self—the ego—rather than the integrated self of Jungian psychology. It's a little confusing, but we can best understand the idea of self-transcendence if we understand it as ego-transcendence.

Transcendence is something we typically associate with sages and mystics, but it happens to ordinary people as well. Achieving transcendence isn't an all-or-nothing proposition. People can experience a little transcendence or a lot. In fact, most people have experienced episodes of transcendence. The state of mind called "flow" is one example. Sometimes called "being in the zone," the flow state occurs when a person is so fully absorbed in what they're doing that they lose their sense of self. The individual merges with the work. No thought is given to the past or future. All that exists is the present moment. One way researchers identify flow-state experiences is by asking respondents to agree or disagree with the statement, "I often become so fascinated with what I'm doing that I get lost in the moment, like I'm detached from time and place."[1] Here we find one of the positive consequences of projection: fascination that dissolves the boundary between the inner and outer worlds.

Flow states may occur when we are engaged in a creative activity or while playing sports, and it brings out the best in us. It may lead to practical benefits, but the flow state doesn't have to produce anything tangible to feel rewarding. Being in a state of flow is intensely pleasurable in itself.

Flow is one manifestation of transcendence, but there are many others. Researchers have developed a questionnaire called the Adult Self-Transcendence Inventory, which quantifies levels of transcendence throughout the life cycle to better understand how transcendence happens and the kinds of growth that occur during its development.[2] The Adult Self-Transcendence Inventory measures five dimensions: self-knowledge, personality integration, peace of mind, non-attachment, and presence in

Sample Items from the Adult Self-Transcendence Inventory

I feel part of something greater than myself.

My happiness is not dependent on other people and things.

I feel compassionate even toward people who have been unkind to me.

I feel that I know myself.

I don't worry about other people's opinions of me.

Material possessions don't mean much to me.

I am accepting of myself, including my faults.

Self-transcendence reflects an acceptance of all aspects of oneself along with a decreasing need to define oneself based on the standards of other people.

the here-and-now. Respondents read a series of statements, and they answer "Yes" or "No" to each one.

When the scale was tested on a group of volunteers who had also been given a standard personality test, their transcendence scores showed a small but significant link to certain personality traits, providing some clues as to what human transcendence may look like.[3] High scores were linked to openness to experience, a trait that includes imagination, aesthetic sensitivity, attentiveness to inner feelings, preference for variety, and intellectual curiosity. High scores were also associated with agreeableness (the tendency to be kind, sympathetic, cooperative, warm, and considerate) as well as conscientiousness (the desire to perform tasks well and to take obligations to others seriously). Lower scores were associated with neuroticism, which is characterized by worry, fear, anger, frustration, envy, jealousy, and guilt.

Although transcendence transforms our inner world, its outer effects may be minimal. There's a Zen proverb that says, "Before enlightenment, chop wood, carry water. After enlightenment, chop wood, carry water." The modern version might be "do the dishes, drive to work," but the point is clear. Many of the changes that occur will be invisible. Nevertheless, they may still be profound. Jung describes transcendence as "a renewal of the personality, working in every direction and penetrating every sphere of life."[4] The mindless task of doing the dishes becomes a few minutes of serenity, a respite from the intellectual efforts that much of modern life requires. Driving to work, which was once a repetitive chore, becomes a new experience every day, because we discover that nothing is ever the same when we're paying full attention to the present moment (one of the dimensions of transcendence). It's like putting on a pair of sunglasses on a glaringly bright day. Everything looks different, while everything stays the same.

In Jungian psychology, the path of transcendence is the unification of the ego and the unconscious. However, before that unification can be accomplished, it's necessary to establish a strong ego that is separate from and acts independently of unconscious instinct. At first glance, that may appear counterproductive. If our goal is the unification of the ego and the unconscious, why would we start by separating them?

As we saw earlier, the ego starts out weak, needing support from parents, teachers, and peers as it undertakes the task of gaining conscious

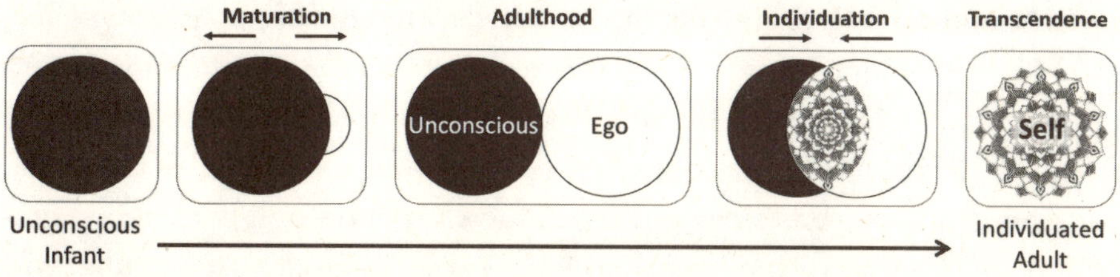

The ego must establish an independent identity before it can join with the unconscious to realize the transcendent self.

control over behavior. As the ego develops, decisions are more often based on reason and planning than on impulsive reactions. It's only after the ego has become strong that it can face the unconscious as an equal. If the ego tries to merge with the unconscious before it has developed its full strength, it runs the risk of being engulfed. Instead of elevating one's animal nature by shining the light of reason into unconscious darkness, the light is snuffed out, and instinct reigns supreme. Recall that when Little Brother couldn't resist drinking from the bewitched stream, he was transformed into an animal.

Achieving unity between the ego and the unconscious by maximizing their differences is a paradox reminiscent of the mystical ten of the Pythagoreans. Whereas the number one is the unconscious ouroboros, ten is transcendent unity in full consciousness: one from many, each of which maintains a unique individuality. We can approximate the idea of unity from differentiation if we think of the partnership between a border collie and its trainer. If you've ever seen a video of a collie herding animals or performing in herding competitions, you've seen two very different creatures, dog and trainer, who have learned to act as a single unit. They don't do it by becoming similar, but by emphasizing their differences. A working border collie isn't dressed in cute outfits and served human food on china plates. Along the same lines, a dog trainer doesn't act on instinct. He doesn't hit his dog when he's angry or sleep in when it's time for morning training. In order to function as a single unit, the dog maximizes its dogginess and the human his humanity. Unlike is joined to unlike, and unity is achieved.

During the maturation phase, the ego acts as the agent of separation, working to differentiate itself from the unconscious. The individuation process, by contrast, occurs via the initiative of the unconscious, which acts as an agent of unification. The transcendent function, that instinct Jung describes as leading to "the revelation of the essential man," is the clearest example of an unconscious agent that pursues unification, but it is not the only one. All the instincts by nature push upward into the light of expression and awareness. As we saw in "The Frog Prince," the drive for unification with the ego can be so insistent that the best metaphors to describe it may be erotic desire and romantic love.

When the fully developed ego begins to incorporate elements of the unconscious, for example by accepting parts of the shadow as legitimate components of the self or balancing rational deliberation with irrational instinct, the person is able to develop transcendent qualities. We can get a sense of the scope of these qualities by looking at the topics covered in the Adult Self-Transcendence Inventory (self-knowledge, personality integration, peace of mind, non-attachment, and presence in the here-and-now), but a fuller understanding of transcendence requires us to look beyond questionnaires. Rating scales are developed by scientists, who test them thoroughly to ensure the scores produced by the scale will be scientifically valid, thereby producing results that reflect real-world behaviors and experiences. That's an advantage of the scientific approach. A disadvantage is that rating scales may not capture the full richness of the human experience. Philosophy and literature can be more helpful in that respect.

French occultist Éliphas Lévi gives us his interpretation of transcendence in his description of the *Magus*. The word derives from the same root as *magic* and is associated with wisdom and philosophy, as we see in the appellation of the three wise men, the Magi, who visit Jesus at his birth. Lévi's description corresponds to some of the items in the Adult Self-Transcendence Inventory, but it goes deeper and is more evocative:

> The Magus welcomes pleasure, accepts wealth, deserves honour, but is never the slave of one of them; he knows how to be poor, to abstain, and to suffer; he endures oblivion willingly because he is lord of his own happiness, and expects or fears nothing from the caprice of fortune. He can love without being beloved; he can create imperishable treasures, and exalt himself above the level of honours or the prizes of the

> lottery. He possesses that which he seeks, namely, profound peace . . . [H]e is a child with children, joyous with the young, staid with the old, patient with the foolish, happy with the wise. He smiles with all who smile, and mourns with all who weep; applauding strength, he is yet indulgent to weakness; offending no one, he has himself no need to pardon, for he never thinks himself offended; he pities those who misconceive him, and seeks an opportunity to serve them; by the force of kindness only does he avenge himself on the ungrateful.[5]

THE TRANSCENDENT FUNCTION

Healthy childhood development happens spontaneously. Toddlers don't need to be told they must learn to walk, and adolescents don't need to intentionally adopt the task of developing abstract reasoning. The process is encoded in their DNA and expresses itself without the child's knowledge or consent. The same is true of individuation in the adult under the influence of the transcendent function. As Jung writes, "[The transcendent function] is in the first place a purely natural process, which may in some cases pursue its course without the knowledge or assistance of the individual, and can sometimes forcibly accomplish itself in the face of opposition."[6] The drive toward transcendence is a biological imperative.

Researchers at Uppsala University in Sweden sent a self-transcendence survey to 2,000 Swedes, ages twenty to eighty-five years, to evaluate the changes that take place as people age.[7] They found that the acquisition of transcendent qualities doesn't proceed at a steady pace. Gains in transcendence occurred in response to events that disrupted existing ways of life. Large increases occurred in women between the ages of twenty and thirty-five, which the researchers interpreted as a response to the challenges of childbirth and motherhood, because the growth curve among childless women was almost identical to that of men. Men experienced large increases between the ages of forty-five and fifty-five, corresponding to the midlife crisis, when they must reevaluate their identity. Some middle-aged men buy flashy sports cars, some have an affair, and others develop a higher humanity. Despite the stereotypes, it appears the last is not uncommon.

When the researchers drilled down to the individual level, they found that younger respondents who had experienced crises scored higher than those whose lives up to that point had been relatively smooth. The events that were most efficacious in increasing transcendence were those that heightened a person's sense of mortality or vulnerability, such as surviving cancer. It's not unusual for people undergoing cancer treatment to make profound changes in their lives. Many switch to a healthier diet and begin exercising regularly. They gain a greater appreciation for their loved ones and the simple pleasures in life. Some ultimately say the illness was a blessing in disguise. Since transcendence is the unification of opposites, we shouldn't be surprised to find that, in this study, the path to peace, self-acceptance, and satisfaction was one of hardship and suffering.

Not every survey respondent benefited from hardship, though. Greater transcendence occurred only among those who had navigated their crises successfully, without feeling crushed by the weight of adversity. Nietzsche's famous aphorism, "What does not kill me makes me stronger," seems to be only half true. Well-developed ego strength is a prerequisite.

Ego strength varies from person to person and usually grows over the course of a lifetime. Some people are lucky; they easily acquire ego strength thanks to a fortunate genetic disposition, and they're able to bounce back quickly after a crisis. People born with less natural ego strength may manage crises successfully through the help of others. Their helpers may provide emotional and material support, serve as role models, or act as auxiliary egos, as the king did for his daughter in "The Frog Prince" when she resisted fulfilling her promise. There are times, however, when crises hit so hard, they overwhelm our personal and social resources and our only option is suffering and perseverance. The path that leads to transcendence can be a hard road to travel, but only the hardest roads lead to the highest heights.

The Swedish survey found that after the midlife crisis, transcendence scores plateaued, followed by another steep climb upward during old age. As we enter our late sixties and beyond, we experience a series of losses. Retirement strips us of an important part of our identity. The body starts to break down. Friends die, and our own death looms larger on the horizon. To a maintain a meaningful existence, a shift in perspective must take place. We have to move away from a materialist, pragmatic approach to life toward a more cosmic, transcendent one.

The researchers saw several recurrent themes among the oldest people in their survey—those above the age of seventy-three—who also had high transcendence scores. They were less self-occupied and felt a sense of being part of a larger whole. They transcended time, feeling the past and present were both available to them at once. Fifty-two percent agreed with the statement, "Today (as compared to when I was 50) I feel the distance between past and present is disappearing." One man spoke of experiencing his memories in great detail: "It is almost as if I am there again. It is like I can go back in time and relive parts of my childhood." The events of childhood were as real as the present, and many were able to reinterpret those events from a mature perspective, a reframing that led to acceptance of difficulties they had suffered. One woman described finding forgiveness for her mother who had mistreated her. Another said, "When I was young, I felt like an isolated lonely little spot in the universe, but nowadays I feel like a link in a generational chain, where the chain itself is the important thing, not me."

Most of the older respondents agreed with the statement, "Today, I take myself less seriously than earlier." Women, in particular, spoke about freedom from obsessing over personal beauty. Their obsessions were replaced with a happy acceptance of their body. One man who had always worn a suit began going out to lunch in sweatpants, a t-shirt, and a baseball cap. He commented, "I don't need to impress anyone anymore." This same man took an interest in his family he had never shown before. His daughter found him playing with Barbies with one of his granddaughters, and he had resumed building model cars, which had been a hobby of his when he was a boy. He said, "At my age, I'm going to act how I want to act."

Another aspect of taking oneself less seriously that the researchers documented was a shift from focusing on one's own needs to focusing on the needs of others, particularly children and grandchildren. The authors of the study wrote, "Egoism has been overshadowed by altruism." They noted that some individuals in the study started giving away their belongings when they reached old age, having found that "the journey through life is easier and more joyful if one is carrying light luggage." They had discovered for themselves the insight of Lao Tzu: "To attain wisdom, remove things every day."

Last, respondents reported a greater acceptance of uncertainty and the mystery of life. Study participants seventy-four and older were more likely to withhold judgments and more reluctant to give advice. They became more tolerant and broad-minded, transcending the proper–improper duality of the inflexible.

All these changes were accompanied by greater life satisfaction. Among those with the highest self-transcendence scores, 50 percent described themselves as very satisfied with their lives, compared to only 8 percent in the lowest bracket. Self-transcendence is a prize worth pursuing.

How, then, do we achieve transcendence? Is it only brought about by the passage of time or endurance of hardship, or are there ways the ego can speed up the process by forging a closer relationship with its unconscious partner?

LISTENING TO THE UNCONSCIOUS

Just as fairy tale characters must listen to and trust their magical helpers before they can accomplish the tasks that will win them their kingdoms, building a strong partnership with the unconscious starts by listening to what it's trying to tell you. It's different from the kind of listening we're used to, because the unconscious generally doesn't use words to express itself. It makes itself known in other ways. Think back one more time to the poet who "look'd up in perfect silence at the stars." Something is being communicated, but it's not possible to put it into words. And yet, the value of the communication is not diminished by its message being indescribable. How do we make sense of these silent communications? How can we decipher what they mean if there are no words to interpret?

Silent communications from the unconscious make themselves known in the form of gut feelings, intuition, and emotions. Sometimes, we understand these communications in a way that's not so different from how we understand language. On the simplest level, language communicates information. Someone might say to us, "I don't think that man's telling the truth." Or, we might have a gut feeling a person can't be trusted. Language can also influence our feelings. If someone said, "Emma told me you did a great job with the presentation yesterday," you'd feel a sense of satisfaction. Similarly, your unconscious might make you feel satisfied, although the reason might not be as obvious. A princess in an old fairy tale who has lost her lover thinks one day while walking through the forest, "I was so sad before, and now I'm happy. I feel as if my true betrothed were around here somewhere." In fact, he has been turned into a bird and is perched in a nearby tree. Sometimes we feel happy for no apparent reason because the unconscious is aware of things the ego doesn't see.

Information, regardless of its source, is not always immediately useful. If you're told that riding roller coasters helps people pass kidney stones, there may not be much you can do with that information. Still, it might come in handy someday. The same is true of unconscious communications, especially when we are first learning to listen and are less fluent with its language. A vague emotion, a sense of uneasiness that comes out of the blue, or a sudden memory of the distant past may not seem relevant at the time you experience it, but at some point, it may coalesce into an important insight. These communications can be seeds of transformation that take years to grow. Don't discard them because you can't see the value in them today. The alchemists knew that the starting material used to produce the philosophers' stone "is of cheap price and found everywhere." Foolish fairy tale characters throw away useless pieces of junk that turn out to have been objects of power. The ones who trust their supernatural helpers, even when they can't see the value in their gifts, are the ones who succeed.

The unconscious, with its massive processing power, has a great deal of wisdom to share with the ego. But there's a benefit even greater than knowledge to be gained, from listening to the unconscious. Whenever one person listens to another, it affects the nature of the relationship between them. This effect is most powerful if the listener gives her undivided attention to the speaker, which is more difficult than it sounds.

It's rare to get another person's full attention for more than a moment or two. Charismatic people know this. They give others their undivided attention to make them feel special, and are often described as making others feel like they are the only person in the room. We do this naturally when someone suffers a loss, such as the death of a loved one. It can be painful being with someone who's suffering, because we feel powerless to help them. But, later, they invariably tell us that our simply listening helped them more than they can express in words. If you've ever been comforted in this way, you know the effect it can have.

If we want to understand the unconscious—if we want to make it our friend and ally—the best strategy is to learn how to listen to it. As hard as it is to listen with undivided attention to another person, it's even harder with the unconscious. The unconscious is intimidating because, you'll recall, it is alien to the ego. Its perspective balances the ego's, which is enormously valuable, but from the ego's point of view, this complementary perspective feels hostile and oppositional. The unconscious is also irrational. Its perspective comes from a place where the opposites are united, which, to the non-transcendent ego, feels like chaos. The unconscious upsets the orderly world of the ego and forces it to confront (or flee) frightening, primitive drives.

Consequently, in order to make the unconscious conscious and move forward along the path toward individuation, the first task is to strengthen the ego so that it can listen with its full attention and not become fatigued while safely facing the archetypal forces that emerge. Additionally, the ego must learn to set aside its bias toward understanding through *logos*—language and reason—and open itself up to nonverbal, experiential forms of communication. The question is how best to accomplish these two tasks—and avoid the dangers of succumbing to the unconscious in the process.

JOURNEYING DOWNWARD

Encounters with the unconscious often involve a lowered state of consciousness, as one might experience in a trance, while sleeping, or in the dreamy state just before falling asleep. Under these circumstances, consciousness is suppressed, and the unconscious plays a more prominent role in our mental life. French psychologist Pierre Janet called this state *abaissement du niveau mental.* Concentration and attention are reduced, along with a loosening of inhibitions and restraints. We don't experience this *abaissement* only when we're sleepy; it also occurs when we merge with the group psyche at a sporting event or a political rally. Nor does it happen only in extraordinary circumstances. We spend approximately half our waking lives in a minor state of *abaissement du niveau mental*. It's called mind-wandering.

Our minds wander whenever we're doing one thing but thinking about something else. People think about work while taking a shower, replay conversations while driving, and think about running errands while playing with their children. In these moments, we operate on autopilot: largely unaware of our surroundings and missing out on the flow of life occurring around us.

To learn more about the effects of mind-wandering, a team of researchers from Harvard University recruited 2,250 volunteers and gave them an app that sent alerts randomly throughout the day.[8] Whenever the volunteers received an alert, they reported on what they were doing, and whether they were thinking about something other than the task at hand. They also reported how happy they felt. The researchers found that mind-wandering occurred about half the time their volunteers were awake, and that people were less happy when their minds were wandering compared to when they were focused on the task at hand. They concluded, "A human mind is a wandering mind, and a wandering mind is an unhappy mind."

The researchers cautioned that mind-wandering isn't all bad. When unconscious processing is predominant, we sometimes get insights into problems we haven't been able to solve with our conscious efforts. Many people get their best ideas in the shower. The problem is that, most of the time, we don't remember the thoughts we have while our minds are wandering. It's not uncommon for a person to emerge from a reverie and have no idea what he was thinking about. Marijuana smokers often experience profound insights while high, but when they come back to reality, they find those "insights" were nothing but clichés.

Going down into the unconscious is one thing. Bringing the treasure found there back into the light of ordinary consciousness is something quite different. A common theme in mythology involves the hero finding a magic talisman and then losing it before he can bring it home. In the Epic of Gilgamesh, a Mesopotamian poem written in the second millennium BCE, the hero brings up from the depths of the sea a miraculous plant that restores youth. But it's stolen by a serpent while he's bathing, and he returns home empty handed. There are fairy tales from many cultures that describe a hero who marries a magical princess and lives happily for many years in her enchanted kingdom. But, when he decides to pay a visit to his former home, to return to his ordinary life, he loses all memory of the princess and her magical world. This plot device is found in the Brothers Grimm fairy tale "The Drummer," the Celtic fairy tale "Smallhead and the King's Son," and the Russian fairy tale "The Water King and Vasilissa the Wise."

To get the most out of a connection with the unconscious mind, we must avoid sinking into a state of lowered consciousness, as we do in sleep, or when our minds wander. We must be wide awake while remaining open to the influence of the unconscious. It's another type of transcendence, one that carries us beyond the duality of alert versus unconscious.

Although the unconscious can submerge conscious awareness, there are rare occasions when it does the opposite, making the ego more alert and thereby raising consciousness to a higher level. A heightened state of consciousness occurs during encounters with the numinous, as we saw in part one. People feel wide awake, and the experience may feel more real than ordinary reality. The flow state is another example of feeling wide awake under the influence of the unconscious. These spontaneous states of transcendence are rare, though, and they can't be called up at will. To consciously encounter the unconscious in a more consistent way, we have to look to other approaches.

Jung described a technique he called active imagination, which allowed him to converse with archetypal characters in his unconscious while fully awake.[9] In some ways, active imagination is similar to Sigmund Freud's free association. The practitioner closes his eyes and allows his imagination to run free. Images or thoughts that appear must be accepted as they are, without any conscious censoring or elaboration. The ego must try to avoid getting caught up in the images—to avoid thinking *about* them, but rather only observe what spontaneously appears. The next step is different from free association, however, and lies at the heart of why the technique is called "active." The practitioner engages the figures he encounters in conversation and allows them to speak, making an effort to avoid putting words in their mouths and simply listening.

Jung's experience with active imagination began with a dream. A white bird flew down and alighted on an emerald table in front of him. It transformed into a woman, who spoke mysterious words to him about "the twelve dead" before turning back into a bird and flying away. In a later dream, he encountered twelve corpses, men from past centuries lying in sarcophagi, who came to life as he gazed on each one—his living consciousness bringing to life the ancient figures within the unconscious.

Trying to understand these dreams led Jung to a possession that was more than metaphorical. "I had no choice but to write everything down in the style selected by the unconscious itself," Jung writes in his autobiographical *Memories, Dreams, Reflections*. "Sometimes it was as if I were hearing it with my ears, sometimes feeling it with my mouth, as if my tongue were formulating words . . . I knew that I had to let myself plummet down into them, as it were. I felt not only violent resistance to this, but a distinct fear. For I was afraid of losing command of myself and becoming a prey to the fantasies."

His work with active imagination began in earnest when he allowed himself to "plummet down" and experienced a waking, fairy-tale vision of a dark, underground world, covered in a soft, sticky mass. He met a dwarf, a gigantic black scarab, a snake, and other mythological creatures. Jung spoke to one of the characters he met, an archetype of the wise old man, who explained to him the objective independence of the unconscious. "He said I treated thoughts as if I generated them myself, but in his view thoughts were like animals in the forest, or people in a room, or birds in the air."

Jung had conversations with other archetypal figures through this practice, and there were occasions when this work made him feel as if he were losing touch with reality. Even when he wasn't engaged in active imagination, he felt that the air around him was filled with ghostly entities who spoke to him. At one point, a voice warned him that failure to discover the secret of what was happening to him would result in suicide. He kept at it only because he felt that, as a doctor, he owed it to his patients to understand these terrible forces within the psyche, even at the cost of his sanity. Under their influence, he abandoned his academic career, leaving the university where he had lectured for almost a decade. For the next three years, he could do no work. He couldn't even read a scientific paper. He ultimately came back to himself by drawing mandalas, ancient symbols of the psychic wholeness of the number four.

Barbara Hannah, a Jungian psychologist, has written that active imagination isn't dangerous when it's done right, but it's hard to do right and easy to do wrong.[10] Active imagination can lead to states of psychosis, especially when there is no one to ground the explorer and keep him in touch with reality.[11] Anyone who attempts active imagination will have their own, unique experience, but it's a dangerous undertaking. It can be a reckless, headlong rush into the depths of the psyche, where the archetypes can easily overwhelm the ego. For most people, navigating the path of transcendence will require a different approach.

Transcendence is the highest state of being, bringing peace, contentment, and compassion. It's a difficult state to achieve, because the ego must give up a certain amount of autonomy and acknowledge the validity of the perspective of its dark twin. The ego must learn to appreciate unconscious ways of understanding the world, and accept even what seems repulsive as part of its own self. Often the ego resists, and, in many cases, the realization of transcendence requires hardship, failure, and pain. Perhaps transcendence is worth the suffering, but it would be nice if there were another path.

In the following chapter, the book's last, we'll explore meditation as one possible path. Meditation is a gradual approach to forming connections with the unconscious rather than a frontal assault on the fortifications that divide the two parts of the mind. It teaches us to hear the messages that rise up from the unconscious, while building our ego strength so we don't get swept away by the archetypal forces found there. It's one of the most direct ways of increasing ego strength, learning to hear the whispers of the unconscious, and cultivating the universal love and compassion that comes with self-transcendence.

THE *CIRCULATIO* AND THE *CONJUNCTIO*

If attention is directed to the unconscious, the unconscious will yield up its contents, and these in turn will fructify the conscious like a fountain of living water.

C. G. JUNG

THE PARADOX

Human beings are alchemists who practice the art upon themselves. Meditation is one form of that art, a psychological manifestation of the alchemical *circulatio* and *conjunctio*.

The essence of meditation is awareness, the most basic characteristic of consciousness. In fact, to be aware and to be conscious are essentially the same thing. Awareness is also the first step in coming to terms with the unconscious. We can go through life believing that our ego is the only agent within our minds. We can do our best to ignore or explain away psychological phenomena that are contrary to the will of the ego or beyond its ability to produce. Or we can cultivate awareness of unconscious influences. We can be on the lookout for states of metaphorical possession and try to understand the gods and goddesses within us, the personifications of human instinct.

Meditation is the practice of increasing one's capacity for conscious awareness, which opens the door to join with these unconscious forces as equal partners, and, in the process, make use of their gifts and discover the nature of our true identity.

Meditation in the West is generally viewed as an activity that leads to better health, like running or eating a healthy diet. That's a reasonable way to think about it. A study that evaluated 600,000 members of a health insurance plan found that long-term meditators had lower medical utilization rates in all categories of care.[1] But meditation began as a mystical exercise intended to lead to enlightenment. The first written description of meditation is found in the Hindu Vedas from around 1500 BCE, but the practice itself is thought to be thousands of years older. It was conceived as a way to achieve self-knowledge and union with the Absolute. Paralleling the imagery of alchemy, the Vedas compared the meditator's mind to refined gold, "pure, bright, soft, workable, and luminous."[2]

In his introduction to the Vedic religion, William Mahony writes, "The practice of meditation as understood in Vedic India was to bring the seeker to an awareness of the eternal and unchanging World-Soul and its essential unity with being itself."[3] The goal of Vedic meditation, like the goal of the Jungian process of individuation, is to achieve the unified self. The Upanishads, Vedic texts of Hindu philosophy, tell us, "He who knows the Self is wise, immortal, self-existent, full of fresh sap, fully complete, free of desire. Knowing the wise, never-aging, eternally youthful Self, one has no fear of death."[4]

The subject of meditation wasn't prominent in Jung's writings, but he was aware of the parallels between his discoveries and Eastern philosophy. In an introduction to a contemporary translation of *The Secret of the Golden Flower*, an ancient Taoist text, Jung writes, "When I began my life-work in the practice of psychiatry and psychotherapy, I was completely ignorant of Chinese philosophy, and only later did my professional experience show me that in my technique I had been unconsciously led along that secret way which has been the preoccupation of the best minds of the East for centuries."[5]

Rather than as a religious practice or an exercise that leads to better health, for our purposes, meditation is most usefully viewed as one of Jung's "secret ways"—as a path toward individuation.

When a person is meditating, it may look like she's in a trance. However, the opposite is true. People in a trance are highly suggestible. The ego that judges and decides is offline; people respond to suggestions automatically, without thought, and when they return to an ordinary state of consciousness, they often remember nothing of what happened. They're functioning at a lower level of consciousness. When a person is meditating, by contrast, she's wide awake and focused. Consciousness is heightened rather than lowered. And yet, the ego is silent.

This paradoxical state of being wide awake without conscious thought is difficult to achieve and requires daily practice. Beginners often fall asleep while trying to meditate. Suppressing the chatter of the conscious ego has the effect of making them sink down into a lowered state of consciousness. The meditator seeks to avoid this *abaissement* by achieving balance. Ego dominance is opposed by abstaining from conscious thought. Unconscious dominance is opposed by staying alert.

Here, we're once again reminded of the mystical ten. In this case, the ego attains harmony with its opposite by being most unlike it—that is, fully conscious. Meditation is an instance of the alchemical marriage of complementary opposites. As it is during the numinous experience and the flow state, in meditation the ego is more awake rather than less.

The importance of being fully conscious is emphasized in other spiritual traditions, as well. In the book of Revelation, God says, "See, I am coming like a thief! Blessed is the one who stays awake." The parable of the ten bridesmaids found in the Gospel of Matthew illustrates the importance of being alert specifically as a preparation for the *conjunctio* (the unification of complementary opposites). The bridegroom is delayed, and everyone falls asleep waiting for him. When he arrives at midnight, everyone scrambles to get ready, but only those who are prepared are brought to the wedding feast. Johann Sebastian Bach based one of his most popular cantatas on this parable; its title is "Sleepers, Awake!"

HOW MEDITATION STRENGTHENS THE EGO

Meditation has been studied for millennia and plays a central role in many religions. It would be impossible to do justice to all that's been written about meditation in only a handful of pages, yet the discipline itself is, in some ways, straightforward. One common practice, called "focused attention," involves forming an intention to pay attention to one thing. If the mind wanders, it's gently brought back to its intended focus. That's it. It sounds simple, but it has profound effects on the mind.

Let's begin with the concept of intention. Forming intentions is an important function of the conscious mind. When we first examined the relationship between the ego and the unconscious, we looked at a woman who was supposed to skip breakfast because she needed a fasting blood test. When she experienced an urge to eat a bagel, she resisted it because it was contrary to her intention. The hunger that came from her unconscious was opposed by the ego.

But it doesn't always work that way. If you recall, that was her second attempt. The previous day, her intention had been frustrated. She had accidently eaten a muffin because her unconscious made her forget the blood draw, hunger beating out rational decision-making.

The reason we can't always choose what we do is because, as we've seen, the ego isn't the only agent that has a say in the matter. The ego doesn't truly have free will, because it plays a limited role in determining behavior. Just as our will may clash with the outside world (you can't drive your car to the store if it has a flat tire), it can also clash with the inner world. When an alcoholic goes into a bar, he may think he can choose to drink only a single beer, but it's unlikely to happen that way. He'll almost certainly follow up that first drink with many others. Even those of us who are free from addiction don't always do what we intend. The spirit is willing, but the flesh is weak.

Although we don't have free will, we do have free intention. We can form an intention to do whatever we want. If we fail, it doesn't change the intention. We can fail once, twice, a hundred times, and still maintain the intention. We pick ourselves up, dust ourselves off, and try again. It's a manifestation of the alchemical *circulatio*.

A meditator forms an intention to pay attention to one thing, often the breath. But when she tries, she finds that her mind wanders. She learns she doesn't have free will. But she puts her failure behind her, and refocuses on her intention, gently bringing her attention back to her breath. Again, her mind wanders; she becomes aware of it, and brings her attention back to her breath. Rinse and repeat, dozens of times during even a brief meditation session: focus attention on the breath, think about something else, bring the attention back to the breath.

A number of things happen as a result of this exercise. The first is that she gets used to failure. At the beginning, it's frustrating, but after a while, after it's happened hundreds of times, it doesn't bother her anymore. Failure is what happens when you try something new. The second thing that happens is she gets good at putting her failure behind her and trying again. Over time, she gradually gets better at paying attention to her breath. She's gaining control over her unruly thoughts. Her ability to turn intention into action grows. In some ways, it's like lifting weights to build muscle. Every time she brings her attention back to her breath, it's like a bodybuilder doing a lift. Over time the muscle (or ego) gets a little stronger, a little better at doing what's being practiced.

When scientists scan the brains of meditators, they see different parts of the brain turning on and off as the meditators shift between attention to their breath and mind-wandering.[6] Forming conscious intentions activates the prefrontal cortex. When the mind wanders, the default mode network is activated. This network involves a collection of areas that are farther back in the brain, along the sides, and deeper below the surface. Every time a meditator notices her mind is wandering and shifts her attention back to her breathing, the prefrontal cortex lights up, and the default mode network becomes quiet. Like the biceps of a weightlifter doing curls, the layers of cells that make up the prefrontal cortex measurably thicken as people become more practiced at meditation.[7]

Another important thing that happens during meditation is that meditators get in the habit of noticing what their mind is doing. Beginning meditators will engage in mind wandering for a long time before they realize they're doing it. Over time, they get better at noticing when it happens, so they can return their attention to the breath sooner.

Monitoring mental activity in this way allows a person to gain perspective on his thoughts. It's like taking a step back and putting some distance between one's thoughts and oneself. "Oh look, my mind is thinking about work. That's not what I want. I'll stop thinking about work and bring my attention back to my breath." Developing the habit of looking at oneself from this objective viewpoint cultivates what's been called "an observing ego." The ego doesn't identify with the thoughts; it takes a step back and observes them objectively, like watching the behavior of "animals in the forest or birds in the air," as Jung's archetypal guide explained to him. Cultivating an observing ego not only makes it easier to notice when your mind is wandering; it also makes it easier to tolerate upsetting thoughts, the kind of thoughts that emerge from the shadow. If we can observe them dispassionately, it becomes easier to accept them and work with them.

Disturbing thoughts trigger anxiety and activate a brain structure called the amygdala, which processes fear and anxiety. The amygdala can be activated by outside threats, such as a snarling dog, or by inner sensations, like physical pain or anxiety-inducing thoughts. When experienced meditators are presented with a threat, either internal or external, they show less activity in the amygdala than novices, suggesting greater tolerance to upsetting emotional experiences. Having the capacity to tolerate the emotional distress caused by an encounter with a powerful and unfamiliar agent from within the unconscious is a necessary first step in coming to terms with it.

When a meditator notices her mind is wandering, she manages it in a specific way. She gently brings her attention back to her breath. *Gently* is an important word. When we fall short of our own expectations, we tend to be harsh with ourselves. If we close a document on our computer without saving it, we think it's fine to call ourselves unkind names, like idiot or loser, even though we wouldn't do the same to someone else. Who—or what—is the target of this verbal abuse? Not the ego. We didn't consciously choose to close the document without saving it. We're attacking an agent in the unconscious, and this creates a hostile, adversarial relationship with it. Meditation can help us stop this bad habit. When a meditator experiences failure, she doesn't condemn the part of her that made her fail, she just tries again.

Gently. Over time, she learns to compassionately accept her weaknesses and shortcomings. By accepting them, she can work with them and move closer to understanding them.

Being gentle with an unconscious agent that opposes the intentions of the ego may allow us to understand its perspective and see the value in its viewpoint. If you've ever had to rewrite something because you accidently closed a document without saving it, you may have ended up with something that was better than the original. Maybe the original didn't meet the standards of the agent who caused you to delete it. Maybe it contained something you were unaware of that would have substantially diminished its value, or it was too one-sided, neglecting an important perspective. Sometimes, things the ego values need to be destroyed, so something better can take their place.

This circular process of paying attention to the breath, experiencing mind-wandering, and gently returning to the breath doesn't sound like much, but it shares aspects with ancient initiation ceremonies, which ritualistically transformed children into adults. Both are designed to strengthen consciousness so it can act independently of instinct. Initiation ceremonies often required initiates to stay awake for extended periods, forcing them to demonstrate their ability to resist the body's natural desire to sink into sleep. In many ceremonies, initiates had to refrain from eating for days to prove their ability to resist instinctual hunger. Some also had to overcome fear and intentionally expose themselves to real or apparent danger, and they voluntarily chose to endure pain.

There are echoes of these heroic achievements of the will in the simple act of noticing the mind is wandering and gently returning the attention to the breath. In both cases the ego asserts its independence and grows stronger. Initiation ceremonies derive their power from the intensity of the experience. With meditation, the power comes from repetition. Each act of will is tiny, but thousands of repetitions allow the effects to build over time, gradually reshaping the meditator's brain.

Let's look at a hypothetical example of how this acquisition of ego strength might help someone manage a difficult situation. Imagine a porcelain collector visiting a friend who shares his interest. His friend hands him an exquisite vase he just acquired, and the man is filled with envy. He doesn't know what to do with this unpleasant feeling. He can't tuck the vase under his arm and run out the front door. There seems to be nothing he can do, and besides, he doesn't like to think of himself as an envious person. It lowers his opinion of himself, so he represses the feeling. Then he drops the vase, and it smashes on the floor.

This is a dramatic example, but things like it really do happen, and a dramatic example can make the forces at play easier to see. The chain of events began when the emotion of envy was passed up to the man's ego, indicating that all was not well within the unconscious. Is there anything the ego could have done to help the unconscious? And if so, what? If the ego wants to be helpful, it needs to do the same thing we would do for a friend who is suffering: listen. When loved ones come to us with their troubles, they're rarely looking for practical advice, and may become annoyed if we try to solve the problem for them. Rather, they need someone to listen, because, as we've noted, listening can comfort and heal in powerful ways. But that's not how our porcelain collector reacted. His unconscious needed him to be a good friend, but instead, he approached the problem in practical terms, which appeared to limit him to either acting on the envy and stealing the vase, or not acting on it and ignoring the emotion.

The man's purely practical approach trapped him in a duality, and since he obviously couldn't steal the vase, he repressed the envy. This was the rational thing to do. Envy doesn't help us in any way, so why should we bother experiencing it? Why should we let this destructive, ugly emotion have its way within our psyche? Unconscious contents are described as "drives" for a reason. They pressure us to express them. When the man denied the envy access to his conscious mind by refusing to listen, it came out as an action, instead. All in all, it was a bad way to manage the problem.

Now let's look at how the practice of meditation might give the man the tools he needed to transcend the duality. A meditator gets in the habit of noticing thoughts and feelings, viewing them objectively from a safe distance, so to speak. The meditator would have quickly noticed the envy, taken a step back from it, and looked at the emotion dispassionately. The man did the opposite. He identified with the shameful emotion and then felt bad about having it.

The meditator wouldn't feel ashamed of his envy, because he's learned that his thoughts and emotions aren't under his conscious control. The meditator is also helped by the fact that his amygdala functions more calmly; the alarm signals that go off in response to upsetting experiences are more subdued. Since the envy doesn't provoke feelings of shame or intolerable distress, the meditator doesn't have to suppress the feeling. In fact, he does the opposite. Developing the habit of paying attention to his thoughts and feelings has made him more curious about who he is. Although the envy isn't a product of his ego, it's the product of something inside him. He wants to get to know that better. He wants to follow the exhortation of the Delphic oracle to "Know thyself." He puts his curiosity into action by turning his attention to the envy and thinking about it. He lifts it up into the light of consciousness.

It may sound odd, but the best thing to do in this situation, the best way to bring the instinctual emotion into a closer relationship with the ego and humanize it, is to experience it the way a wine connoisseur savors a glass of fine Bordeaux, exploring the subtleties of the emotion. Someone practiced in self-observation might try to discover how his envy over the vase is different from every other feeling of envy he's had in the past. He might make the unconscious conscious using *logos* by describing to himself in words the unique features of the emotion. Later, once he is no longer with his friend, he might take the use of *logos* a step further by giving the words a physical reality through talking about the emotion with someone else or writing about it in a journal.

It might seem as if immersing oneself in envy would increase the discomfort produced by this unpleasant emotion. That may happen at first, but most people are surprised to find how quickly the sting fades away. If you flee from an unpleasant emotion, it will haunt you. Your imagination will blow it up into larger-than-life proportions. On the other hand, if you gaze at the emotion with steady eyes, it quickly shrinks down to a manageable size, and can be tolerated easily. Just like in the movies, the monster you can't see is always scarier than the one you can. Facing things head on, putting the feelings into words—even when they're the most frightening aspects of ourselves—is the easiest way to subdue their unruly nature.

If the man had responded to his envy by telling his friend what he admired about the vase, perhaps attempting to add details to what his friend already knew, his conscious envy might lend enthusiasm to his ego's efforts to recall and describe relevant facts. His envy would consciously express itself in the form of excitement and eloquence.

Acceptance is a good strategy for managing many kinds of unwelcome experiences. Millions of Americans live with chronic pain, which, like emotion, is the product of unconscious, bottom-up processing; the neural signals we interpret as pain flow from the spinal cord and deep brain structures up into the prefrontal cortex. Research has shown that, as with envy, managing pain with acceptance is the best way to avoid its disruptive consequences.[8]

An analysis of the way people respond to pain found that two strategies were linked to lower levels of distress and disability.[9] The first was willingness to experience the pain. Chronic pain patients were presented with a series of statements and asked to record which ones they agreed with and which ones they didn't. Endorsing the statement, "I would gladly sacrifice important things in my life to control this pain better" was linked to more distress and more disability than agreeing with one that said, "Controlling my pain is less important than any other goals in my life." The difference couldn't be accounted for by pain intensity alone, suggesting that acceptance was playing an important role.

The second strategy was choosing to engage in life activities regardless of pain. Someone using this strategy would disagree with the statement, "I avoid putting myself in situations where my pain might increase." They would be more likely to endorse the statement, "I am getting on with the business of living no matter what my level of pain is."

The role of the ego in managing physical pain is the same as it is in managing emotional pain. The unpleasant sensation is acknowledged, accepted, and even welcomed, if possible. No attempt is made to suppress it, but it's not allowed to dictate choices and behaviors. Actions remain under the control of the rational part of the psyche, which decides that life will be lived to the fullest despite the pain. This maintenance of control is made easier by acceptance. People who fight their pain—those who believe they must find a way to control it before they can live a normal life—often end up living in fear. They restrict the scope of their activities and think about their pain more than anything else. Taking an adversarial stance against the unconscious drains the ego of its resources.

The ego and the unconscious are so different from each other that we often reflexively choose the adversarial approach. We assume the familiar way is the right way. Pain is unpleasant. The logical thing to do is look for ways to get rid of it. The same is true of envy and other products of the shadow. The ego wants its house to be clean and orderly, and it tries to clear away whatever doesn't fit. But the ego has a roommate it must learn to accommodate. It's best to think carefully before initiating hostilities. A more sensible approach is to begin guiding these aspects in our unconscious from the shadow into the light.

THE TRANSFORMATIONAL POWER OF *LOGOS*

By making us more aware of our mental activity, meditation can help make the unconscious conscious, which can lead to a transformation of the unconscious contents. When something in the unconscious moves into consciousness, we often describe ourselves as "realizing" something: "I just realized why that person makes me so uncomfortable." We speak in this way because when we make something conscious, we make it real. When *logos* translates unconscious processes into words, it names them, and doing so gives them both reality and meaning. Naming is the first step in taming the primitive forces of the unconscious.

Naming things is one of humanity's essential roles. In Genesis, God brought all the animals to Adam to see what he would call them. Jung writes that he believed consciousness was necessary because "after thousands and millions of years, someone had to realize that this wonderful world of mountains and oceans, suns and moons, galaxies and nebulae, plants and animals, *exists*."[10] The role of consciousness is to acknowledge, and through conscious acknowledgment, the instinctual becomes intelligible.

In folklore, knowing someone or something's true name gives a person power over it. In the fairy tale "Rumpelstiltskin," a girl who has made a promise to a magical creature frees herself by learning its name. In Judaism, a miracle-working rabbi is called Baal Shem, which is Hebrew for "Master of Names," and in Taoist traditions, knowing the true name of spirits and demons gives one power of them.[11] For the human brain, naming is similar to personification. Names make things intelligible by allowing the ego to conceptualize them—by serving as a handle for the conscious mind to grasp. As Jung writes, "The *magically powerful name* gets a grip on the object. This not only renders the object harmless, but incorporates it into the psychic system, thus increasing the meaning and power of the human mind."[12]

When something is *nameless*, by contrast, it is unknowable to the ego, and the term is often used to designate something dark, powerful, and evil. In Tolkien's *The Lord of the Rings*, the wizard Gandalf says, "Far, far below the deepest delving of the Dwarves, the world is gnawed by nameless things . . . Now I have walked there, but I will bring no report to darken the light of day."[13]

Giving something a name can raise it above its natural origins and humanize it. In the children's book series *Moomintroll*, author Tove Jansson describes a tiny creature who lived its whole life without a name, until one day, it meets its hero, who names it Teety-woo. The creature says, "You see, before I had a name I just used to hop around, and everything was simply happening around me, sometimes nice things, and sometimes not nice, but nothing was real, don't you see? Now I'm a person, and everything that happens *means* something. Because it doesn't only happen, it happens to *me*, Teety-woo. And Teety-woo may think this or that about it, as the case may be—if you see what I mean?"[14] Undifferentiated, unconscious behaviors gain an identity when they're given a name. They acquire intelligible meaning and become more human. It's part of the process of making them *praus*.

For most people, envy is part of their shadow. By naming and acknowledging a piece of the shadow with gentleness and acceptance, we tame it. We don't express it in its raw form. If the porcelain collector had chosen not to repress his desire to steal the vase, and instead made that desire conscious, he would have brought it into the light and made it an ally. And the excitement and eloquence it gave him would have enabled him to increase his friend's enjoyment of the latest addition to his collection.

It isn't complicated. All we have to do is listen to and accept our unconscious impulses as part of ourselves. But it is difficult. Those impulses were unconscious for a reason. They force us to confront the darker aspects of ourselves, making us realize we're not as civilized as we may have thought. These instinctual forces of nature are also powerful. If we're not prepared, they easily overwhelm us. In order to maintain control, we cannot allow ourselves to succumb to a state of *abaissement du niveau mental*. We must be wide awake enough to face the strength of our lower passions.

Fortunately, even the most difficult, potentially overwhelming emotions are easier to face when the ego has been strengthened through meditation. In a PTSD treatment study, veterans with severe, combat-related PTSD were given a course of treatment that included weekly meditation sessions over three months. At the end, most veterans had milder symptoms, and the number who had dropped out of treatment was lower than in similar groups of veterans treated using other techniques.[15] If the effects of meditation are powerful enough to help veterans face the trauma of war, we can be confident meditation will give us the strength we need to face our inner demons.

MINDFULNESS

The first type of meditation—focusing on a single thing, such as the breath—teaches the ego to be both wide awake and silent at the same time. It enhances the ego's ability to pay attention to mental events without becoming engulfed by them. Once one has achieved this ability to focus on the breath, the next step is to move on to what's called "open monitoring" or "mindfulness." Mindfulness is described as a *receptive practice* because there's no single object of focus. The mind is receptive to all forms of physical and mental experience.

In ordinary terms, to be mindful is to pay attention to one's surroundings. In psychology, mindful awareness means paying attention in a way that's nonjudgmental, nonreactive, and focused on the here and now. It also involves something called "meta-awareness," that is, being aware that you're aware. There's a difference between being angry and being aware that you're angry. In the first case, it's easy to identify with the anger and be swept away by it. But when we are aware of our anger, it brings a perspective that permits separation from the emotion, opening up space for us to make a rational decision about how to best respond to it.

When applied to the physical environment, mindful awareness is being open to the sensations of smell, taste, touch, sound, and vision. Attention is paid to how they continually change, as you immerse yourself in the experiential world of becoming. As the Zen proverb says, "When walking, walk. When eating, eat." An effort is made to withhold judgment, to experience the sensations in the present moment, without getting caught up in whether they're good or bad. You might look up and see the sky covered in dark clouds. An ordinary reaction would be to worry that it's going to rain. A mindful approach, by contrast, would be to take in the majesty of the storm clouds, smell the air, and be aware of the overall atmosphere of the impending thunderstorm. The intention of the mindful approach is to allow experiential processing to act unimpeded, and as a result, non-ordinary states of consciousness occur more readily, such as experiencing the *genius loci* of the thunderstorm hovering in the air.

We're so used to the automatic running commentary inside our heads that stopping the flow of these thoughts to focus exclusively outward may sound like we're pursuing a kind of mental emptiness. But the silent, focused mind can be experienced as quite full, just in a new way. A Tibetan book of meditation describes it this way: "The nature of the settled mind is a clear, vivid brilliance, not a total nothingness. In this settled state there is clear, open, resplendent, gently flowing consciousness."[16] Marie-Louise von Franz compares it to the philosophers' stone: "The stone symbolizes a form of consciousness that is just pure being, beyond emotions, fantasies, feelings, and the current of thoughts that characterizes ego consciousness. It is a unity that merely exists, that was and is always there, unchanging."[17]

In addition to awareness of physical sensations, mindfulness is also awareness of mental processes—thoughts, feelings, and emotions—as they occur in the present moment. Similar to physical sensations, these internal events are experienced in a nonjudgmental way, like a meditator accepting his envy. Emotions are part of nature just as much as wind and rain. That understanding allows the person to take a step back from the emotion and accept it as an authentic part of one's self without suffering guilt, a distracting emotion that makes it harder to think clearly.

Although mindfulness is nonjudgmental and nonreactive, it's not passive. As with focused attention on the breath, giving bare attention to sensations, thoughts, and feelings requires effort. In fact, many people are surprised at how difficult it is to simply give spontaneous phenomena—both inner and outer—one's full attention. The ego has to lift itself out of its semiconscious mind-wanderings, like a swimmer treading water to keep her head above the waves. In the beginning, it's hard to maintain for more than a few minutes.

Although being mindful can be challenging under ordinary circumstances, we naturally and effortlessly become mindful when we're in danger. Life-and-death situations encourage the ego and the unconscious to work together. They put aside their differences, at least for a while, to maximize the odds of survival. Walking down the street, you might be lost in a semiconscious fog of worry, anticipation, or just random thoughts, but if someone on an electric scooter almost runs you over, your mind instantly snaps into focus as your attention wakes up to the challenges of the present moment.

Danger's ability to activate the unconscious in a way that fills the ego with energy may account for some of the attraction of extreme sports. Flying a hang glider off the edge of a cliff requires total concentration. A single moment of distraction can lead to tragedy, so the unconscious gives the ego the energy it needs to be fully aware in the present, doing all it can to keep you alive. Veteran skydivers who were asked to explain why they engaged in such a risky recreational activity described feelings of acute well-being, peace, calm, and stillness, as well as ecstasy, immortality, and altered perceptions of time and space.[18] Normally, achieving such a heightened state of consciousness would require decades of practice.

Extreme sports are among the most reliable ways of bringing oneself into the present moment, but almost any danger will do. As English author Samuel Johnson said in reference to an Anglican clergyman who had been sentenced to execution, "When a man knows he is to be hanged in a fortnight, it concentrates his mind wonderfully."

It's one thing to live transcendently in the present moment as one jumps out of an airplane; it's another to maintain conscious awareness for long periods. Conscious awareness requires a great deal of mental energy, because when we're mentally fatigued, we tend to drop into lower levels of consciousness. As we've noted, the ego has little energy of its own, so it must rely on the ouroboros, the self-renewing principle of fertility within the psyche. The unconscious operates on its own terms, independent of the wishes of the ego, and when we're faced with a hard task—such as maintaining conscious awareness or finding an answer to a seemingly impossible dilemma—we can only hope that it will provide us with the mental resources we need. However, that doesn't mean there is nothing we can do to maximize the chance our unconscious will come to our aid.

In fairy tales, a character setting out on a long journey sometimes gets lucky, and an animal comes along and gives him a ride. Foolish characters may try to capture the animal, tie it up so it can't escape. That never works out well. In real life it would be like using performance-enhancing drugs to boost mental and physical energy. When these drugs are used as medications to treat illnesses, they generally have positive effects. But when people use them to squeeze additional performance out of themselves—to chemically force their brains to perform the way the ego wants—the brain rebels. Sometimes the drugs just stop working. The magic beast slips free of the ropes that bind it. Other times, misuse of these drugs ends in catastrophe. The animal turns on its master and attacks him.

By contrast, a wise fairy tale character makes friends with the animal. He makes sacrifices for it. He learns the animal's needs and does all he can to help it. Sometimes he postpones his own quest to help a poor creature he meets along the way. But it always works out to his advantage. As we saw in "The Golden Bird," the fox's friendship gives the hero the wisdom and strength he lacks, life-saving help that's available to him whenever he's in danger.

That's what mindfulness (and meditation, more generally) does: befriends the instinctual unconscious, and, through repetition, consistency, and gentle guidance, tames it, to forge a working relationship. In the beginning, we have to use willpower to stay alert while meditating, just like a beginner tennis player needs to intentionally move her arm to carry out her swing. But if it's done enough times, the experiential system takes over, and conscious resources are no longer needed. Once that happens, the tennis player can execute a good swing whenever she needs to, and the meditator can maintain a high level of conscious awareness without becoming fatigued. A small piece of the ouroboros has been made *praus*. It requires an enormous amount of hard, sustained work and a sacrifice of time. But the power of the unconscious is so much greater than anything the ego has at its disposal that the prize is worth the effort.

THE DANGERS OF MEDITATION

We saw in part one that encounters with the unconscious are not without risk, especially if we're in a state of *abaissement du niveau mental*, when the ego is unprepared for the archetypal invasion. However, because of its emphasis on an active, wide-awake ego, meditation is a safer way of opening the lines of communication with the unconscious. Instinctual forces are less likely to overwhelm an ego that has its wits about it. But that's not always the case. In one survey, about 25 percent of regular meditators reported disturbing experiences, including feelings of fear, dread, and terror.[19] Usually these feelings were transient and didn't require interruption of the meditation practice, but in some cases, meditation can result in psychosis, mania, and suicidality. Practices that strengthen ego discipline, such as focused attention on the breath, were less likely to trigger negative reactions compared to open monitoring, in which meditators used experiential processing to give spontaneous thoughts and feelings full access to conscious awareness.

The survey found that female meditators were less likely to have disturbing experiences than men. Religious people, who may be more comfortable with non-ordinary states of consciousness, were also less likely to have bad experiences. The risk of meditators having negative experiences went up during retreats, in which people engage in long, intensive meditation sessions while isolating themselves from the outside world. Many retreat organizers ask that people being treated for a mental illness receive clearance from their doctors before participating.

In firsthand descriptions of unpleasant meditation experiences collected by researchers, one participant reported a "feeling of falling into a void." Another wrote, "During a meditation a painful image came to my mind, and my heart started to beat fast, and I started to feel fear and panic." Opening a connection to the unconscious mind can release mental energy. Usually this energy can be used by the ego, allowing it to more effectively pursue its goals. But not always. One meditator described "feelings of being over-energized and an inability to sleep." Some meditators became so consumed with the practice they felt "the time not spent meditating is wasted."[20] Travelers who enter the land of Faerie sometimes have trouble finding their way home.

Take the case of a thirtysomething man seeking treatment for bipolar disorder who was an avid meditator. He was good-humored, personable, warm, and demonstrated a commitment to spiritual growth. He tried to become a monk but had been asked to leave the monastery because of bizarre behavior that included sneaking into the kitchen in the middle of the night to steal food. He couldn't explain why he did it, and it appeared these episodes occurred in an altered state of mind. It was almost as if he were sleepwalking, similar to a strange side effect some people experience after taking the sleeping pill Ambien. Such patients get out of bed at night and binge on food, but have no memory of it the next morning. The feeding instinct acts autonomously.

Despite having to leave the monastery, the young man continued to meditate regularly with no apparent difficulties. However, when he attended an intensive, five-day meditation retreat, he became psychotic. After recovering from the episode, he agreed that he shouldn't attend any more of these retreats, but he couldn't stop. In some ways he was like a man addicted to drugs. He committed himself to abstinence, but then, after a while, the desire built up to the point at which he could no longer resist it, and he lost sight of the dangers. A few months later, he attempted to reenter the monastery, and when that failed, he joined a meditation commune, where he sank deeper into psychosis. Like the songs of the Sirens that drive sailors mad with desire, the allure of diving down into his unconscious proved to be his undoing.

It's usually best to allow the darkness inside ourselves to become known—to make the unconscious conscious, even in its ugliest aspects—despite the less pleasant effects that result. At the same time, we need to be careful. Meditation should never be practiced under the influence of intoxicating drugs, which break down the boundary between ego and unconscious (unless you're a shaman who's been trained for this kind of work from childhood). Even in the absence of intoxication, people who have experienced psychosis should discuss meditation with their doctor before engaging in the practice, because their ego–unconscious boundary has been weakened by an illness that adversely affects brain functioning. The dangers inherent in growth sometimes can't be avoided, but by treading carefully, starting slow, and strengthening the ego as we gradually open the door to the influences of the unconscious, the dangers can be minimized.

LOVING KINDNESS

After focused attention and open monitoring, the third type of meditation is loving kindness, which draws upon the predisposition of the unconscious to connect and unite, bringing us closer to the mystical goal of universal love.

Loving kindness is most closely associated with Buddhism, but the call for universal love and compassion is also found in the pre-Buddhist Upanishads of Hinduism and in the early sutras of Jainism. In the Western tradition, Pythagoras is quoted as saying, "When each person loves the other as much as himself, it makes one out of many."[21] In the Christian Bible, Jesus tells his disciples that to love God and your neighbor are the two highest commandments.

Determining the principles that should guide our behavior toward others is the goal of the study of ethics. It's a discipline that's usually pursued using the tools of logic and reason, but acting ethically often requires more than logic and reason. The problem with a rational approach is that, while it tells us *what* to do, it doesn't necessarily help us do it.

It's easy to decide to be a good and loving person when you're sitting comfortably in an armchair. But when the time comes to act, you may not do what you intend. Knowing and doing are different things. One doesn't naturally follow the other. Although the ego excels at understanding theoretical concepts, to put those concepts into action, the ego has to use willpower, which is a poor tool for the job. Willpower fatigues too easily and is frequently overcome by impulses from the unconscious. After waiting on hold for ten minutes and being transferred to three different customer service representatives, you may intend to be kind to the person who picks up the phone, but there are stronger forces at work.

For our actions to correspond with our intentions, we need the help of the unconscious. Bottom-up experiential processing can make ethical behavior effortless. But it's also resistant to change. If we want ethical decision-making to be fast, automatic, and effortless, we'll need to dedicate ourselves to hours of training, just as we would when learning to hit a tennis ball. Aristotle of course wasn't aware of dual processing theory, but he did know it was hard to make good decisions in the heat of the moment. He writes that in order to consistently practice virtue, a person needed to cultivate an ethical disposition through habituation.[22] A theoretical understanding isn't enough; developing good habits requires time, effort, and repetition.

We can train the experiential processing system to make ethical behavior intuitive and effortless with a kind of meditation called *metta,* the Pali word for "loving-kindness." Metta meditation has some things in common with focused attention. The meditator gives his full attention to a series of benevolent wishes, starting with himself and expanding outward. He begins with, "May I be happy." Whenever his mind drifts off, he brings his attention back to "May I be happy," focusing on his wish for happiness. He then wishes, "May I be healthy. May I be peaceful. May I be loved," or other, similar phrases. After spending time directing loving kindness toward himself, his next step is to direct it toward a loved one, by picturing that person and focusing his attention on "May you be happy. May you be healthy. May you be peaceful. May you be loved." Next, he directs loving kindness toward a casual acquaintance, then someone he sees as an enemy, and finally the whole world.

A wish is an act of will. The meditator doesn't think *about* loving kindness, he *engages* in loving kindness. It's an important distinction, like the difference between thinking about skiing and getting on the slopes. One will make you a better skier; the other probably won't. Wishing plays a prominent role in magic. For example, the modern ritual of blowing out the candles on a birthday cake is accompanied by a wish. There are echoes of alchemy and the worship of celestial bodies in the common expression "to wish upon a star." Wishes are often found in fairy tales. Although none of the characters in "The Frog Prince" engages in wishing, the reader is alerted that this will be a story about magic with the opening words, "In old times when wishing still helped one . . ."

When we perform an action over and over again, such as wishing for health, happiness, and peace, the neural circuits that code for that action become stronger and easier to activate. It's like making a path through a forest. Each time you walk along a particular route, the path becomes more distinct and easier to follow. When a commuter steps up to the turnstile at a subway station, she reaches into her pocket and takes out her fare card without giving it conscious thought. The goal of metta is to do the same thing with ethics—to make universal benevolence the default, so the meditator automatically acts with loving kindness.

Initially, the ego sees this repetitive work as mindless, even distasteful. Thinking the same thing over and over may feel like brainwashing yourself. Aristotle's insistence that virtue must be cultivated as a habit through repetition has been criticized as downright unphilosophical![23] But if we want to transform thinking into doing—swords into pentacles, to use the symbolism of the tarot—we have to accept that the unconscious experiential system works differently, and we have to engage it on its own terms. The animal we're trying to make *praus* learns in a different way than the rational ego.

Metta begins with compassion for oneself. That may feel a little self-indulgent, but many people adopt a harsh, contemptuous attitude toward themselves, a problem that often begins in childhood. Parental criticism, shaming, and neglect make a child feel judged and insecure. They see themselves as weak, inferior, and unworthy.[24] Self-criticism is mistakenly seen as a safety behavior, in which people attack themselves in order to correct their perceived flaws so that they avoid being attacked by others.[25] But self-criticism is harder to defend against, because it comes from the inside. It leads to feelings of powerlessness and can make a person anxious and depressed. Patients being treated for clinical depression are less likely to respond to both psychotherapy and antidepressant medication if they have high levels of self-criticism.[26] It's also important to remember that when we condemn ourselves, we're usually condemning our unconscious, the characteristics that lie outside conscious control. Feelings of worthlessness are generally less about intellectual capabilities and more about the deeper aspects of the personality, fostering a state of hostile conflict between the two parts of the psyche. Loving-kindness meditation breaks down the old habit of self-criticism, reduces negative feelings, and increases positive emotions. It can even bring about long-term changes in personality.[27]

Even people who have a less critical view of themselves benefit from self-compassion. Medieval theologian Thomas Aquinas saw it as a prerequisite for loving others. As he writes in the *Summa Theologica*, "The love with which a man loves himself is the form and root of friendship. For if we have friendship with others it is because we do unto them as we do unto ourselves."[28]

Once we have learned compassion toward ourselves, we can turn our attention to compassion toward others. The ability to notice what's happening in one's own mind increases the ability to recognize the state of mind of others, which in turn enhances the capacity for "putting yourself in someone else's shoes"—the essence of compassion. The word *compassion* derives from Latin, meaning "to suffer together." When we discussed the power of listening, we noted that when people are suffering, they usually don't get much comfort from others telling them how to solve their problems. Rather, they need someone to witness their suffering, to see it through their eyes and suffer along with them, sharing and lightening the load.

A quiet amygdala allows the expert meditator to better tolerate distress, so it's easier to be present with those who are suffering. However, while the amygdala becomes less active in response to others' emotions, other parts of the brain become more active. To put it another way, the brains of meditators react more to the experiences of others, but those reactions don't upset them. When researchers compared brain scans of novice and expert meditators, they found that parts of the experts' brains associated with compassion and sympathy had stronger reactions to the sound of both a child laughing and a child crying for help.[29] Like Lévi's Magus, the expert meditators had trained their neural pathways to smile with all who smile and mourn with all who weep.

Even a short-term course of loving-kindness meditation can change the way a person responds to the suffering of others.[30] After only an eight-week program, research volunteers were five times more likely to give up their seat to a man on crutches compared to other volunteers who didn't participate in the loving-kindness meditation course. A third group of volunteers, who took part in an eight-week course of mindfulness meditation, also failed to experience a boost in compassion. It appears that metta has specific effects and needs to be practiced along with the other two forms of meditation to achieve this kind of growth.

Loving-kindness meditation has the ability to transform the way we interact with others and may even change the way we understand our place in the cosmos. Universal love is the most comprehensive expression of the alchemical *conjunctio*, creating unity from multiplicity.

RATIONAL MYSTICISM

When the ego acknowledges that the unconscious processes information in a way that's beyond its understanding, while also making an essential contribution to how well the psyche performs, it practices a kind of rational mysticism.

The rational ego requires a great deal of fuel—not only physical energy, but also desire, to replace the need for willpower. The rational ego looks at its own limitations and realizes that a closer partnership with the hidden personalities inside the psyche will help it achieve its goals. But in order to make the unconscious conscious, and thereby gain access to its energy, the ego must learn to think in a different way than it's accustomed to. If your partner speaks a foreign language, it's rational to learn that language, even if it appears nonsensical at first. And the language of the unconscious—of experiential processing—is mysticism.

Bringing the ego and the unconscious together requires work. Most of that work comes from the ouroboros. Unconscious contents drive themselves upward with the goal of being expressed. In animals, that expression takes the form of instinctual behavior. Humans engage in instinctual behavior, too, but we have a second option. Unconscious contents can be expressed by being made conscious, thereby avoiding the need to enact destructive impulses. Raising the unconscious up into the light happens naturally. The unconscious provides the energy; the ego just needs to pay attention.

However, the ego needs to pay attention in a special way. It must allow the unconscious drives to occupy its awareness without identifying with them: to feel anger without being angry, to feel pride without being proud. It takes strength to keep the required distance, and the ego needs to be prepared. If you're going to let in the beasts, training yourself to handle them is a rational thing to do.

There are no shortcuts when it comes to this training. Every quadrant of the labyrinth must be traveled. Mistakes will be made again and again. To climb the hard path of seven, the path of spiritual growth that leads to the transcendent eight, failure must become a familiar friend. The wisest thing to do is embrace the *circulatio*. Train the will in small and gentle ways, over and over. If wrestling with the *circulatio* is what growth requires, the rational thing to do is to step into the ring. You make the most progress when you know how the system works.

One of the challenges of meditation as a path to growth is that the conscious mind rebels against such a simplistic expression of the *circulatio*—doing the same thing, repeating the same words, over and over again. The intellect is symbolized by Mercury, the god of language. He is quicksilver, the reflecting shapeshifter, the fleet-footed runner, fastest of all the gods. The ego seeks answers. It wants to jump to the finish line in a single leap. It's an affront to have to give in to the dull, plodding toad of an unconscious that requires the ego to repeat the same thing ten thousand times.

But that's the way the brain works. The unconscious uses experiential processing, which requires repetition to get good at something. The rational ego may be faster, but its development also requires time and effort. Counting your education and the reading you've done since graduation, you've invested a lot of time developing your conscious mind. That's time well spent, and it's important to continue to stimulate the rational ego throughout life. But it makes no sense to neglect the other side, because it too is critical to success. You have to accept that it's not like you. It doesn't think like you do. Yes, it's a dull, plodding toad, but it's a dull, plodding toad that carries riches the ego cannot imagine.

Science has provided us with enormous wealth. Continued innovation is lifting millions out of poverty and giving developed nations tools to live in more sustainable ways. The answers science provides are satisfying because they make the world comprehensible, giving us control over chaos. And yet, the material benefits that science has provided have not been matched by spiritual benefits. Science doesn't help us become better human beings. It doesn't help us overcome our fears. It doesn't make us joyful or increase our capacity for love. That's not a criticism of science. Those are things outside its domain. But when we lose sight of the boundaries of science and start to believe that everything lies within its power, we lose an important part of our humanity.

Recognizing the value of magic is difficult in this age of scientific preeminence. It's easier to say it's just a matter of time before science gives us everything we need. Visions of the approaching earthly paradise of six tempt us to believe that, thanks to the power of reason and logic, we've outgrown the need for the irrational aspects of the human psyche. But that overlooks the fact that even the best scientists aren't afraid of a little irrationality. As physicist Richard Feynman has said, "If you think you understand quantum mechanics, you don't understand quantum mechanics." It's reasonable to accept unreason, when unreason works.

The unconscious parts of the brain give us art, music, and spirituality; courage, love, and joy. Most importantly, they give us ourselves. If you reject the dark regions of the mind, your quest to become who you are will fail. Your own mind will become a monster you cannot overcome, a cave you dare not enter.

Reason is a great blessing, the gift of an awakened consciousness. But it cannot stand on its own. The higher cannot stand without the lower. Our true humanity won't be achieved until we unite reason with unreason, light with darkness, science with magic. To live life fully, we must become whole.

Appendix

THE PLANETS AND THEIR ALCHEMICAL SIGNIFICANCE

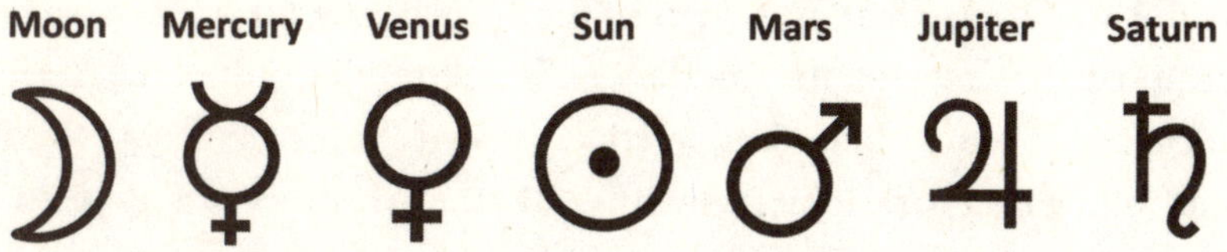

The alchemical symbols for the classical planets.

THE MOON

The Moon was associated with the goddess Luna and the archetypal feminine because of the belief that the monthly cycles of the Moon paralleled the menstrual cycle ("As above, so below"). Because of its effects on the tides, the Moon was also linked to water, and thus emotion. The metal associated with the Moon was the white, lustrous silver. Silver has a propensity to tarnish, connecting it to the phases of the Moon, which waxes and wanes over time rather than always maintaining its full brilliance, like the Sun. The

Moon was called the Queen of the Night, and, as the ruler of darkness, it has become the heavenly body most closely associated with the unconscious as well as the state of being overwhelmed by the unconscious—that is, lunacy.

MERCURY

See page 160.

VENUS

Venus is named for the goddess of fertility, femininity, love, and desire. Her metal was copper, which is relatively soft and has a pinkish-orange color reminiscent of tanned skin (think Coppertone sunscreen). Copper is also the color of grain that's ripe for the harvest. Copper readily reacts with oxygen to form verdigris, the green patina that gives the Statue of Liberty its color. Thus, the metal could represent thriving, green vegetation as well as the golden harvest. Venus as the goddess of love brought one of life's greatest blessings, but love can be a double-edged sword. Sometimes it's life affirming, and other times it's toxic, which is why the word *venom* is derived from the same ancient term as *Venus*.

THE SUN

See page 161.

MARS

The next planet, Mars, is called the Red Planet because it has a reddish color that's visible to the naked eye. It's named after the Roman god of war, possibly because its reddish hue reminded the ancients of blood.

The god Mars could unleash wild frenzy in a warrior, but he was also the patron of martial discipline. An eighth-century BCE Homeric hymn to Ares (the Greek god of war) gives us an example of the planetary god as a psychological force: "Shed down a kindly ray from above . . . that I may be able to drive away bitter cowardice from my head . . . Restrain also the keen fury of my heart which provokes me to tread the ways of blood-curdling strife." The poet asks the god of war to drive away not only cowardice but also fury. Centuries before Aristotle articulated the theory that courage was the mean between the extremes of cowardice and rashness, we find the idea of *praus*, the virtue of being both fierce and disciplined, in the ancient prayer to Ares.

Mars's metal was iron, which was used to craft weapons, and may also have become associated with Mars based on a lucky guess. We know now that the planet's reddish color is caused by rust, iron oxide, which is prevalent on its surface.

JUPITER

Jupiter, whose name is derived from the words *sky* and *father*, is named after the king of the gods. He was also called Jove, and represented royalty, glory, abundance, and also hearty good humor, reflected in the English word *jovial*. Santa Claus is a modern mythical figure who shares some of Jupiter's characteristics.

Jupiter's metal was tin, which at first glance is disappointing. One would think that the king of the gods would rate a nobler metal. However, in ancient times, tin was considered a semi-noble metal and used to make jewelry. It was also associated with the festal characteristics of the jovial god. Because it is nontoxic and resists corrosion, eating utensils, mugs, and plates were made from tin. Tin has a highly crystalline structure, which gives it a pleasing sound when struck, and so bells were also made from this Jovian metal.

SATURN

Saturn was the outermost planet known to the ancients, the end of the line, and therefore a symbol of limitation, whose restraint counterbalances Jovian expansiveness. The god Saturn was portrayed as an old, decrepit man. Saturn is the slowest of the visible planets. Its course through the heavens takes almost thirty years compared to only eighty-eight days for swift-footed Mercury. In Greek mythology, he was Kronos, the god of time, who gives us words such as "chronological" and "chronic." Saturn represented not only things that were old and worn out, but also patience, endurance in the face of hardship, and silent contemplation. Whereas Jupiter was the archetype of feasting and celebration, Saturn was austerity, self-discipline, and asceticism.

Bipolar disorder, also known as manic-depressive illness, is a psychiatric condition in which the archetypes of Jupiter and Saturn play prominent roles. Manic episodes are ruled by Jupiter. The mood of mania is grandiose and expansive. People experiencing mania make big plans, spend great sums of money (often on gifts for others), and pursue pleasure with feverish passion. Depressive episodes, by contrast, fall within Saturn's realm. They're characterized by slow movements, low energy, loss of interest in the outside world, ruminative contemplation, and thoughts of death.

Saturn's metal was dull, heavy lead, the least lustrous of the classical metals. It's also the least resonant. When lead is struck, the vibrations are immediately smothered in dullness. Lead is extremely resistant to corrosion and seems to last forever. Lead pipes constructed by the ancient Romans are still in use today. Corresponding to Saturn's boundary role as the outermost of the classical planets, lead represents a boundary of heaviness. It has the highest atomic number of any stable element.

Lead was considered the basest of all the metals, but it was respected by the alchemists because they believed it contained within itself the power of transformation, like the caterpillar that turns into a butterfly. In its native state, lead is rarely found alone. It's usually found with silver, and often also with gold, leading the ancients to conclude that over time, lead gradually evolved into the nobler metals. One of the goals of the alchemists was to discover ways to speed up this natural process in the laboratory.

THE KEYS OF THE TAROT AND THE PERFECTION OF SOUL

A tarot deck consists of four suits of fourteen cards each (ten number cards and four face cards), plus twenty-two major arcana cards, or keys, which set the tarot apart from more familiar decks. The keys are divided into three groups of seven cards each, representing the three parts of the soul described by Plato in *The Republic*: *eros* (desire or appetite), *thymos* (spiritedness or will), and *logos* (reason). Plato felt it was necessary to divide the soul into parts because an undivided soul could not be in conflict with itself, such as when a person struggles to resist temptation.

According to Plato, people who are primarily driven by the soul of appetite have a love of gain, whether they desire pleasure, possessions, or money. The soul of appetite is perfected by the virtue temperance, in which harmony among the desires (which leads to health) is valued above excess. Plato recommended the study of music as the best way to bring about this harmony. The soul of appetite was believed to reside in the belly.

Those primarily driven by the soul of spiritedness or will have a love of power and prestige. The perfection of the soul of will is fortitude, strength, and courage. Its excess was anger, cruelty, and blind ambition. Plato recommended athletics as the best way to perfect the soul of will, which was thought to reside in the heart. This connection can be seen in English words derived from the Latin word for heart, *cor*, which gives us both *coronary* and also *courage*.

The soul of reason was thought to reside in the head. Its abilities include the use of logic and the capacity to understand abstract concepts. It was perfected through prudence, a word that in modern times can have negative connotations, evoking an excess of caution or, even worse, a puritanical condemnation of all forms of fun. No one today aspires to be a prude, but in earlier times, it had a more positive meaning. For the ancients, prudence meant doing the right thing, at the right time, in the right way, for the right reason. Plato recommended the study of geometry to perfect the soul of reason.

The progression of the keys tells a story of the acquisition of these virtues of temperance, strength, and prudence. It illustrates the trials and tribulations people experience as they come to know their inner strength and use their innate capabilities to overcome the obstacles that stand in the way of becoming fully human.

0: THE FOOL

See page 198.

I: THE MAGICIAN

See page 199.

II: THE HIGH PRIESTESS

See page 201.

III: THE EMPRESS

See page 202.

IV: THE EMPEROR

See page 203.

V: THE HIEROPHANT

The material four of the Emperor is followed by five, the quincunx. Recall that four is a comfortable place of stability. It's a good place to rest. Here, spirit is added to the equation to move us along our journey. In the suits, the fifth card, which takes us out of four, is often a negative one, portraying conflict, suffering, or loss, because the changes required for growth are frequently unwelcome.

At this point on our path, we've mastered our ability to manage the material world. We feel powerful and confident. Now we are forced to look beyond the world of the senses by the Hierophant, Greek for "high priest." The Hierophant, like the High Priestess, opens the door to realms beyond the confines of the ego. But while she is chthonic, oriented downward toward our natural roots, he points upward into the spiritual realm.

Although this is the fifth card (the number of human spirit), three, the prototypical number of spirit, plays a prominent role. The Hierophant wears the triregnum, or triple crown, of the papacy. Three nails project from the top of the crown, representing Christ's passion when he was nailed to the cross. In his left hand he holds the papal cross, a staff with three horizontal bars near the top, and three crosses arranged in a vertical pattern adorn the white band that flows down his robe. There is also a hint of transcendent wholeness in the illustration. At the bottom of the band, a fourth symbol is seen: the four-sided diamond that represents matter and adds balance to the profusion of spiritual symbols.

VI: THE LOVERS

The sixth key is the Lovers, which depicts Adam and Eve in the Garden of Eden. At first glance, it appears to illustrate the six of the earthly paradise. But we've gained the power of the Hierophant, and so the love of the two figures transcends sensuality. They aren't having sex; they're being joined in marriage, a relationship that combines both chthonic and spiritual elements. Here we have two instances of the symbol of three, which takes duality and returns it to unity. There is the marriage itself, which joins two people to form a married couple, and there is the intermingling of the chthonic and the spiritual that combine to form the marital bond.

Lust, the desire for a physical body, is similar to other desires that target things within the material world, such as hunger. But romantic love is different. It generally includes desire for the body of the beloved, but it also involves desire for something incorporeal: the quintessence of the beloved. We have transcended materiality and raised our soul of appetite to a place where it has gained the ability to love an idea. This form of passionate desire will be essential as we move on to the next set of cards, in which we encounter the abstract virtues. It won't be enough to pursue these traits with the intellect only. The road to virtue is too hard. To master them, we must be driven by passion. The higher cannot stand without the lower.

VII: THE CHARIOT

The same passion that drives desire for material goods, the gifts of the Empress, can now be used to drive us forward as we face the challenges of the soul of will. The Chariot takes us to this next level.

Earlier, we looked at Plato's description of the three-part soul in *The Republic*. The Chariot takes its imagery from another Platonic dialogue that discusses the three-part soul, *The Phaedrus*, in which Plato compares the soul to a charioteer driving two winged horses. One is a bedraggled dark horse that is unruly and difficult to control. It represents *eros*, the soul of appetite. The other is a noble horse, light in color, which loves honor. It represents *thymos*, the soul of spiritedness. The white horse tries to ascend upward, but the dark horse tries to drag it down. The charioteer, who must control these two quarreling beasts to reach his goal, is *logos*, the soul of reason.

In the dialogue, Plato appears to be making a value judgment about the disparate parts of the soul, valuing *thymos* over *eros*. But the alchemists remind us that the production of the philosophers' stone begins with "that which is trodden underfoot in the dunghill." We must accept and perfect our lower drives by making them conscious. If we are successful, they become powerful allies, providing us with the life and strength that flow from the instincts.

In the Waite-Smith deck, the charioteer is a warrior, a choice that anticipates the next series of cards, which deal with the soul of will. In earlier decks, the charioteer was Venus, indicating that the motive energy necessary to overcome the challenges we will face on the next stage of our journey comes from love and desire.

VIII: STRENGTH

The first step toward perfecting the soul of will is to develop strength, the subject of the eighth card and an archetype we've encountered a number of times. In the card, we see a woman with a lion. In some ways, this is an unexpected choice. A more conventional depiction of strength might portray the Greek hero Hercules killing the mighty Nemean lion with his bare hands, as he did in the first of his twelve labors. Instead, the card shows a woman gently closing a lion's mouth with a caress, as it licks her hand. The chain around the lion's neck that connects the two is made of flowers. Rather than fighting the lion, she is taming it, making it *praus* so it will be her ally, obeying her commands while retaining its fierce nature.

IX: THE HERMIT

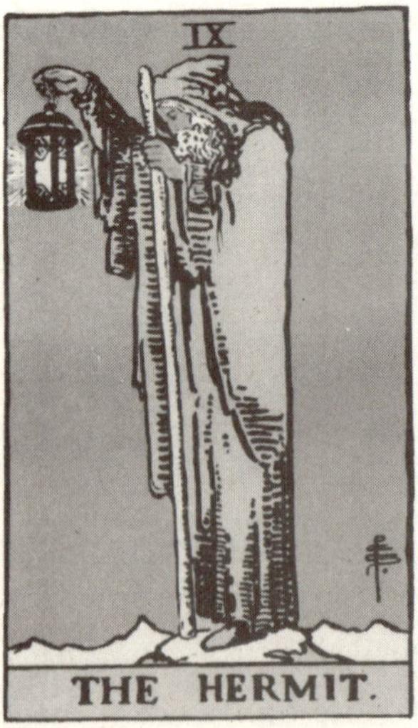

Now that we have gained access to our strength, we are no longer at the mercy of our destructive impulses. The archetype of desire gives us the passion we need to accomplish our goals, but it can also lead to excess. The ascetic Hermit archetype competes with, or we might say balances, the Empress. Hermits give away all their worldly possessions and go into the wilderness to live a solitary life of contemplation, undisturbed by material desires.

The landscape of the Hermit card is bare; there is nothing but the Hermit himself, standing on a snow-covered mountain peak. He lights his way with a lantern, which is illuminated by a six-pointed star, the unification of the upward- and the downward-pointing triangles. This suggests the Hermit's lonely contemplation has given him the ability to see beyond the illusion of apparent duality. At the same time, it also implies a one-sidedness. A symbol of Venus (the intersection of male and female) has been confined within a cage. The lamp illuminates but does not heat.

X: THE WHEEL OF FORTUNE

We have gained our strength and overcome the snares of the sensuous world. We are ready to set out to seek honor and glory. However, life is not so simple, and what we encounter is the Wheel of Fortune, which symbolizes the vagaries of fate.

The expectation of justice, of being treated fairly, is an essential quality of humanity. We expect that hard work will be rewarded, that talent will be acknowledged, and that no one will receive special treatment they don't deserve. But life is unjust, and that can be hard to bear. The Wheel of Fortune symbolizes the cruelty of arbitrary fate.

The wheel has eight spokes, which represent the eight heavenly spheres and their influence on the fate of humans. There are Latin and Hebrew letters around the wheel. The Latin letters, *ROTA*, use the same letters in *tarot* to spell the Latin word for *wheel*. The Hebrew letters are the four letters of the tetragrammaton, an unpronounceable word that spells the mystical name of God, the master of human fate.

The two creatures on either side of the wheel are Hermanubis and Typhon, Hellenized Egyptian deities who represent the forces of good and evil. In *The Pictorial Key to the Tarot*, an illustrated guide to the symbols in the Waite-Smith tarot, A. E. Waite writes that the four corners are occupied by the "four living creatures" of the prophet Ezekiel's vision. Each of the four is described by the prophet as having four faces: "a man's face with a lion's face on the right, a bull's face on the left, and . . . an eagle's face." The Wheel is driven by fearsome forces, against which mortals have little power. We can only hope the wheel doesn't break us.

The Wheel of Fortune from an Italian tarot, made at the turn of the sixteenth century. *(Source: Metropolitan Museum of Art, New York, via Robert Place,* The Tarot: History, Symbolism, and Divination.*)*

At the top of the wheel sits a sphinx, who played an important role in the legend of Oedipus the king. When Oedipus was born, an oracle revealed that he was destined to kill his father. In order to forestall this horrible fate, his parents abandoned him on a mountaintop. But before the infant died, he was rescued by a shepherd, who adopted him, thereby setting the stage for the fulfillment of the prophecy. The sphinx reminds us that we cannot avoid our destiny. Struggling to escape our fate only serves to bring it about.

A deck produced around 1500 takes a somewhat different approach. It shows a man lying on the ground, dreaming about his future. Floating above him is a creature with an animal's head ascending the wheel, saying, "I will reign." At the top of the wheel is a creature who is fully animal, who says, "I reign." Finally, there is a character who is fully human falling down the other side of the wheel, saying "I have reigned." The card suggests that when we seek worldly success, the kind of success the Wheel has to offer, we lose part of our humanity. It isn't success that makes us human, but failure.

Most of us do our best to avoid failure. However, the lessons we learn from failure are among the most valuable. For doctors, failure can be a matter of life and death, and in recognition of this, hospital physicians regularly gather for what's called a morbidity and mortality conference. One of the doctors presents a case that went badly, and the others help identify areas where the medical team made mistakes. The emotions that well up within a doctor when she realizes she's committed an error that has caused suffering or death lock in the memory of what occurred—or to use the symbolism of the tarot, they serve as a very effective cup. Although doctors can't remember everything they learn from lectures and textbooks, they never forget their serious failures. This form of learning is so important that, in most jurisdictions, nothing said in a morbidly and mortality conference can be used against a doctor in a malpractice lawsuit. This enables doctors to speak freely, without fear of legal consequences.

Failure can also make us less judgmental. It opens our eyes to our own imperfections, making us more tolerant of others' imperfections. Failure frees us from arrogance and makes us gentle.

The Wheel is an embodiment of the alchemical *circulatio*. We must go around and around, over and over, until we learn its lesson. Worldly honor and fame are fleeting. They are given and taken away capriciously and deceive us into mistaking good luck for personal merit. In the end, it isn't success that moves us along the difficult path of growth; it's failure.

XI: JUSTICE

Justice is a powerful archetype that has a profound influence over human behavior. People desire justice as one of the greatest goods. They will fight for it and die for it.

Squares, symbols of justice, are prominent in this card. The shape adorns the clasp of Justice's cape as well as the center of her crown, which has squares rather than spikes around its rim. She holds a sword and scales, traditional judicial symbols, but what lies behind her is covered by a veil, symbolizing the unknowable aspects of justice.

Philosophers have written more about justice than any other subject, but they've never been able to agree on what it is. A simplistic definition is that justice is giving every person what they deserve, but that just begs the question: What does any particular individual deserve? Determining that is impossible. Without knowing everything about a person—their upbringing, their inner struggles, the talents and vulnerabilities they were born with—we can't judge that person fairly. And if we can't judge fairly, how will we know what kind of treatment others deserve?

Many spiritual traditions untangle the hopeless knot of justice by stepping away from the rational approach of the ego and turning to the mystical perspective of the unconscious and the aspiration of universal love. Whether a person is a friend or an enemy, deserving or undeserving, it may be that the only way to avoid injustice is to treat them with love.

XII: THE HANGED MAN

We can't judge others fairly, because we can't see into their minds to fully understand their struggles. But there is one mind we are able to see into—our own—and introspection gives us the ability to identify instances of injustice within ourselves. Overcoming those injustices may require a revolutionary shift in the way we view ourselves and the world.

The Hanged Man shows a man suspended upside down by his ankle. It's a card of sacrifice because it's not easy to give up what's become familiar and comfortable. Even when we know we're going down the wrong path, our inclination is to try to make minor course corrections rather than admit failure, throw everything away, and start fresh in a new direction.

Since the Hanged Man represents a reversal in attitude, it's helpful to turn the card over and see him from the opposite perspective. Now we see a man adopting the pose of a dancer. It's a pose of discipline, a virtue we will need in order to overcome the inertia of our current path. The Hanged Man is like the Emperor; he is active, not passive. He chooses radical change. His hair is like flames, and his head is resplendent in glory, shining with the light of consciousness.

XIII: DEATH

Death is the fruit of the Hanged Man's sacrifice. The card depicts a skeletal figure mounted on a pale horse, like the fourth horseman of the apocalypse. A king lies dead at his feet, and in front of him are a child, a young woman, and a bishop, his next victims. Death spares no one.

It seems a little early in our journey through the major arcana for death to be making an appearance, so we can conclude that the death to which the card refers isn't the final death, but rather the partial death that accompanies growth and transformation. This is the death the fox experienced in "The Golden Bird," when it begged the hero to shoot it and chop off its head so it could become human. It's the feeling of loss and emptiness that comes with any sacrifice, great or small. When we make a change, part of one's self dies. The alcoholic who becomes sober is rewarded with good health, but the part of him that craved the pleasures of wild carousing or the comfort of dull senselessness must die, and that can be hard to bear. The bitter water of alchemy is also a kind of death. Once we see with clear eyes, the person we thought we were passes away.

The more we identify ourselves with the thing that must be sacrificed, the greater the feeling of death. Eastern traditions emphasize that the true self is separate from the nonessential things we typically identify with: our possessions, our appearance, even our abilities. In one form of yogic meditation, the practitioner chants, "I am not this body; this body is not mine. I am not this mind; this mind is not mine." Body and mind, whether good or bad, are granted by the Wheel of Fortune. They're separate from the core of one's essential being. Recognizing this truth makes it easier to hand over to death those things that hold us back. Distinguishing the essential from the nonessential makes death an ally, a kind of surgeon who cuts away the corrupted parts of the soul so the healthy parts can flourish.

The card depicts death carrying a flag emblazoned with a white, five-petaled rose, a symbol of consciousness. In order to get the most out of death, we must voluntarily allow corruption to be cut away. If we don't do it voluntarily, the transcendent drive may maneuver us into a place where death acts in spite of our wishes. However, we don't benefit then as much as we do when we adopt the attitude of the Hanged Man, who makes the sacrifice consciously.

Off in the distance we see the sun rising between two towers, the new dawn. In many fairy tales, the old king must die so the young hero can ascend to the throne and usher in a new golden age. Death is not the end; it's a renewal. Death carries us forward, bringing us closer to our goal.

XIV: TEMPERANCE

Temperance is one of the cardinal virtues. In psychological terms, it represents instinctual appetites balanced by rational deliberation, or, on a higher level, the integration of the two. With temperance, the soul of appetite reaches its perfection. It's a higher level than asceticism, which rejects the pleasures of the material world. Temperance allows us to enjoy the gifts of the Empress—to possess them, without them possessing us. The warped desires of greed, indulgence, anger, and blind ambition, which drove us to excess, have been given over to death. Appetite is now in harmony with the

rest of the soul and becomes a force for good, bringing pleasure without loss of control.

The word *temperance* is derived from a Latin word that means to mix in the proper proportion. We see this meaning in the familiar saying, "Temper justice with mercy." Temperance seeks to balance the appetites and establish a harmonious compounding of our spiritual and material natures, symbolized in the illustration by the triangle (spirit) within the square (matter) on the figure's chest.

The card is full of symbols of balance. The winged figure in the center pours liquid from one cup to another, mixing the fluids to achieve the ideal proportion. She dips one of her feet in a stream, while the other rests on dry land, giving her the ability to enjoy the sensuous pleasure of the flowing water without being washed away. At her feet grow irises, a flower named after the goddess of the rainbow, who bridges the gap between the earth and the sky with a ribbon of ethereal beauty. Some decks depict the rainbow itself. It symbolizes the joining of the spiritual archetypes of the sky with the chthonic archetypes of the earth.

No part of the soul can be perfected until all parts are. The soul of appetite couldn't reach its full potential, as it does here, until the soul of will had been developed. The harmony within, which has been achieved through the perfection of the soul of will, can now be reflected in our relationship with the outside world. As within, so without. This idea is illustrated on the card with a strange inversion. In the distance we see a crown rising between two mountains, where the sun should be. In the foreground, the figure's head, instead of being crowned, shines like the sun. The cosmic and the personal are reversed. The microcosm and the macrocosm have become one.

Temperance is the last card in the soul of will series and anticipates the next section, the soul of reason. The harmonious connection between the inner and outer worlds that Temperance gives us sets the stage for reason to rule the instinctual aspects of the psyche with justice. Plato believed reason was the highest faculty of the soul because it permits us to look beyond the material world of becoming into the divine world of being.

Reason is a function of the ego, and we might question the wisdom of privileging the ego above the unconscious. Magical traditions help us understand that much of life is controlled by inner forces that exist beyond the confines of the ego; by placing reason at the pinnacle of human development, are we reverting to the one-sided view of the psyche where we began? Not necessarily. The problem with an exclusive focus on the rational ego, which is characteristic of enlightenment philosophy, is that it's unbalanced. It lacks the harmony that permeates the Temperance card. The goal of individuation is to integrate the various aspects of the psyche by making the unconscious conscious. Each must play its proper role. The ego, with its ability to reason, must govern. Despite its frailty, consciousness elevates the psyche and carries us beyond our natural origins by allowing us to perceive the immutable laws of reason—the eternal world of being. Yet, despite holding a position of superiority, consciousness can't rule as a tyrant. It must learn the lesson of the Strength card and gain ascendency over the powerful instincts of the unconscious through acceptance, gentleness, and love.

XV: THE DEVIL

The gods and goddesses of antiquity no longer enjoy a prominent place in the modern imagination, but the Devil continues to fascinate us. As the personification of evil, he is frequently found in books, movies, and television shows for both children and adults.

In *The Screwtape Letters*, written by C. S. Lewis as a series of letters from a demon to his young nephew, the demon offers advice on how to secure the soul of an intended victim: "Keep everything hazy in his mind now, and you will have all eternity wherein to amuse yourself by producing in him the peculiar kind of clarity which Hell affords." The first task in the development of the soul of reason is to face the clarity of Hell.

The Devil represents those aspects of the shadow that cannot be tamed and must be cast out. When we mastered Strength, the brutality of our animal nature was made *praus*. Temperance allowed us to bring our unruly desires under the sovereignty of consciousness. But we're not done yet. There are worse things than brutish appetites. Hatred, cruelty, and contempt for our fellow human beings are also part of the human condition. To continue down the path of growth, we must first face the vilest of what's inside us.

One of the names of the Devil is Lucifer, "the bringer of light." The card shows the Devil holding a torch, which is pointed downward. The things he illuminates are the lowest depths of the soul. These are the ugliest parts of our psyche, the parts that are most difficult to face. The usual way to manage these horrors is to turn away from them, or to trick ourselves into believing they're not really there. But to enter the highest realm, the realm of *logos*, we must be honest about who we are. We must drink the bitter water.

Despite the pain of having the flames of Hell expose our darkest secrets, few things are more valuable than knowing the true nature of your enemy. We're powerless against forces we can't see. The enemy that is hidden is far more dangerous than the one we can see. The male and female figures on the card have horns and tails, representing regression to a subhuman level, and are chained to a block of stone, signifying the loss of their freedom. It's an alarming condition to find oneself in, but now that the horns, tails, and chains are visible, something can be done about them.

Earlier, we noted that the evil sorceress in Spenser's *The Faerie Queene* was named Duessa, after the number two. What she was and how she appeared were two different things. Corruption dresses itself in beautiful clothes, and too often, we gladly go along with the deception. Our anger is righteous, our vanity is a true reflection of our superiority, and our greed is a blameless desire for what is rightly ours. But when reason opens our eyes to our true nature, the disguises fall away, and what we see is repellent. That gives us an advantage. Freeing ourselves from the chains of our own evil will still be difficult, but now we are motivated by fear and disgust at who we have allowed ourselves to become, and the job gets a little easier.

Will Carroll, the drummer for the thrash metal band Death Angel, said he met Satan in Hell while in a COVID-19 coma.[1] The Devil turned him into a bloated Jabba the Hutt–like monster, and he vomited blood until he had a heart attack, an ordeal he said he wouldn't wish on his worst enemy. But it made him decide to reform his ways, especially his tendency toward self-indulgence. "I don't think Satan's quite as cool as I used to," he added. It's understandable why he would feel that way, but the devil archetype performed its proper role: providing a counterpoint to Carroll's conscious attitude. Although the journey was terrifying, seeing himself with the "peculiar clarity of Hell" moved him forward along the path of growth in a way that perhaps nothing else could have done.

XVI: THE TOWER

We've faced sacrifice, death, and the Devil himself, but there's one more calamity we have to undergo before the light begins to dawn: the Tower. The ego has done everything it can to join forces with the unconscious and make itself strong. It tamed the instinctual desires, found its strength, and identified justice as the proper goal to pursue with that strength. But sometimes the efforts of the ego are not enough. Stubborn obstacles hold us back, aspects of the status quo that are so precious, the ego cannot voluntarily let them go. When that happens, the transcendent function takes over. Whatever continues to block our growth must be destroyed.

The Tower symbolizes the archetype of ruin. Its card depicts a tower being struck by lightning. People are thrown from it and fall headfirst to the ground surrounded by flames. The crown that once sat atop the tower is overthrown.

A tower can be a place of safety, but it can also be a place of imprisonment. Often, it's both. A position of power, the security of wealth, or some other thing we have unwisely crowned as sovereign within our psyche is often the last thing holding us back.

When Rapunzel reached puberty, the witch locked her in a tower and cut her off from the world. It took years for her to escape, and doing so nearly caused the death of her lover. In fairy tales, escape sometimes requires a catastrophe, and the same may be true in real life.

It's not hard to find examples of the Tower, in which individuals fall from the heights of privilege because of apparent careless mistakes or incomprehensibly bad judgment. In November 2019, Texas state representative Alfonso Nevárez was carrying a small envelope containing cocaine when it accidently fell out of his pocket as he was leaving the Austin-Bergstrom International Airport.[2] It was caught on surveillance video, and Nevárez was arrested. He faced up to ten years in prison, but he said, "In a weird way I am grateful. Grief and addiction were consuming me, but oddly enough, I feel better now than I have in a long time."

The Tower can accomplish transformation within our lives in a way that nothing else can, but it's a terrible experience to go through. It may feel as if the world is falling apart. Sometimes, it's obvious what to do. Nevárez entered treatment. At other times, we may be utterly disoriented and have no idea how to pick up the pieces and move forward. The flames in the card's illustration, instead of rising up like they should, fall downward from the sky like rain. The world has been turned upside down; everything is backward, and nothing makes sense. Chaos rules. As Jungian scholar Marie-Louise von Franz writes, "In such moments all well-meant, sensible advice is completely useless . . . There's only one thing that seems to work; and that is to turn directly toward the approaching darkness without prejudice and totally naively, and to try to find out what its secret aim is and what it wants from you."[3]

XVII: THE STAR

Our eyes are now ready to receive the light of reason. The next three cards represent heavenly bodies of increasing brilliance. We are approaching the source of light, clarity, and truth. The Star is the first of the celestial lights; it is the quiet after the storm that shook the Tower to its foundations. Seven white octagrams symbolize the ladder of the seven planets, which the soul climbs to reach perfection. But with this card, our climb is over; the upward striving has come to an end. We have reached the eighth level, the peace of the unchanging sphere of the fixed stars. We tasted the bitterness of death, faced the devil within us, and saw our world crumble to pieces around us. But we persevered. Everything nonessential has been stripped away, and all that remains is that which is imperishable.

The purified soul is portrayed as a nude figure pouring water on the land and into the sea, gently washing away any remaining impurities. Waite tells us she expresses eternal youth and beauty, and that her water is the Water of Life, which she pours out freely. One foot is in the water, while her knee rests on dry land. She unites the opposites of wet and dry. She is the frog of fairy tales, now visible in her true form. The ugly toad is revealed as a star.

XVIII: THE MOON

The Moon card shows the moon rising between two towers. These are the same two towers we saw in the Death card, but, as artist and tarot scholar Robert Place points out, on that card the towers were seen off in the distance, across a body of water. In this card, we've crossed to the other side. It was a terrible crossing, but now it's over, and we've reached a place of calm.

The phases of the moon determine the months of the year in the old lunar calendar, making the moon a symbol of time. The use of the moon to mark time dates back to the Stone Age,[4] and it's still used today by Hindus, Jews,

Buddhists, Christians, and Muslims to determine religious holidays. In the foreground we see a crustacean, described by Waite as "that which comes out of the depths." The journey across the water has returned us to an earlier stage of evolution, an invertebrate animal, and it will take time for us to grow back into our human state. The Moon is a card of rest and preparation.

The moon is the mistress of the night and symbolizes the unconscious. This is the only key that contains no human figures, only animals. Before we can take our final step, that of complete individuation in the sunshine of full consciousness, we must solidify our connection to our instinctual roots. The higher cannot stand without the lower. If we develop a one-sided attachment to consciousness, if we reject our lower nature and try to "climb without steps," as the alchemist Morienus warns, we will fall, and all our work will be wasted.

How long must we wait? Sleeping Beauty slept for a hundred years before the deadly thorns blocking the way to the castle were transformed into flowers that allowed the prince to pass. We may have to wait a long time. However, the Moon is a card of passivity; we don't have to struggle, only be patient. Waite tells us that the message of the card is "Peace, be still."

XIX: THE SUN

The wait is over. The instinctual animal of the Moon card has evolved into a newborn child on a white horse, holding a flag of victory. The white horse symbolizes animal nature reborn into a state of innocence. In the background the sun shines in its full glory. We are no longer limited to seeing by the reflected light of the Moon. We have become strong, and our way is guided by the source of light itself: the Sun. We have reached full consciousness, and we are ready to become whole.

Waite tells us that the child symbolizes simplicity. He describes it as being both "Nature and Art," writing, "When the self-knowing spirit has dawned in the consciousness above the natural mind, that mind in its renewal leads forth the animal nature in a state of perfect conformity." The unconscious can only be perfected when it's fully illuminated by consciousness. When that light dawns, and the two sides of the psyche learn to accommodate each other in perfect harmony, we experience renewal.

XX: JUDGEMENT

At last our eyes are fully open, and we see the truth. The Judgement card depicts the dead rising from their graves to be judged by God. The figures are naked because nothing is hidden. Their arms are outstretched in joy. Robert Place notes that the woman is on the right and the man is on left, a reversal of the usual arrangement of masculine and feminine. We are seeing things with new eyes, from a new perspective.

Judgement in the soul of reason series complements Justice in the soul of will series. Earlier, we developed an intention to act with justice, but discerning the just from the unjust was beyond our abilities. Now we see clearly, not only with the light of reason, but also with the intuition of instinct. This brings us to the greatest virtue of all, the one that encompasses all the others: prudence. Now that the hidden forces within the unconscious have been raised into consciousness, our judgment is honest, unbiased, and clear. There is no longer projection, because we have become conscious of all that was hidden. All our faculties work together in harmony so that we are able to do the right thing, at the right time, in the right way, for the right reason. We have become just.

XXI: THE WORLD

The final key, our ultimate goal, is the *anima mundi*, the world soul, which represents the mystical connection among all living things. As Plato writes, "The world is indeed a living being endowed with a soul and intelligence . . . a single visible living entity containing all other living entities, which by their nature are all related."[5]

From a Jungian perspective, the *anima mundi* represents all aspects of the psyche brought into harmony. It's the personal version of the cosmic ten of the Pythagoreans. The World card depicts the world soul as a beautiful nude, dancing within a laurel wreath, a sign of victory. She is the quincunx, the pure essence, surrounded by the symbols of the four evangelists, Matthew, Mark, Luke, and John (man, lion, ox, and eagle). The wreath is reminiscent of the ouroboros, but rather than engulfing her in the darkness of unconscious nature, it marks her as victorious. She has not defeated her animal nature, nor has she succumbed to it; she has made it her own. The wands she holds match the Magician's from the first card, but he had only one, which he gripped tightly with his right hand. Here, the soul carries two, lightly held, one in each hand. She has achieved balance.

Waite writes, "[The card] represents also the perfection and end of the Cosmos, the secret which is within it, the rapture of the universe when it understands itself in God." We have heeded the admonition of the Delphic oracle. We know ourselves.

ACKNOWLEDGMENTS

The path this book took from conceptualization to completion was not always smooth, and I'm grateful to the many people who helped me along the way.

My son Sam was the one who got things started. The many conversations we had about magic kindled my excitement for the subject, and prompted me to look into my old books by Carl Jung, which had led me to a career in psychiatry many years ago. It took more than two years for me to figure out the best way to approach this topic, and Michael Long, my coauthor for *The Molecule of More*, generously helped me along the way. He spent many hours reading and commenting, even rewriting sections to their great benefit. Thanks are also due for the valuable advice and comments from my colleague Jacqueline Posada. I'm indebted to her for the title of this book.

My wonderful agents, Wendy Levinson and Andrea Somberg, were unwaveringly kind and supportive as I struggled to find the right way to communicate the ideas in this book.

Many thanks to all the talented people at BenBella Books. In particular, I want to thank Jennifer Canzoneri for her timely encouragement when I needed it most, and Glenn Yeffeth for his guidance on the best way to approach the subject. I owe an enormous debt of gratitude to Leah Wilson, who edited this book. The many hours she dedicated to the manuscript brought a scattered collection of ideas into an understandable, unified, flowing narrative. Thanks also to copy editor James Fraleigh, who has a sharp eye, a pitch-perfect ear for good prose, and a seemingly inexhaustible fund of knowledge of everything under the sun.

Most of all, I'd like to thank my wife, Masami, for her unfailing support and encouragement. She believed in the book from the beginning and wouldn't allow me to give it up when I despaired of being able to bring it to fruition.

ENDNOTES

INTO THE DARKNESS

1 Dijksterhuis, Ap, and Loran F. Nordgren. "A Theory of Unconscious Thought." *Perspectives on Psychological Science* 1, no. 2 (2006): 95–109.

If you're interested in learning more about the scientific studies cited in this book, type the title of the paper into https://scholar.google.com/ and click the result. Occasionally, you'll get the whole paper, but usually you'll be shown a summary, called an abstract, of the study's methods and conclusions.

2 Ibid.

3 Jung, C. G. *Collected Works of C. G. Jung, Vol. 8: Structure & Dynamics of the Psyche*. Princeton, NJ: Princeton University Press, 2014.

4 Singer, Wolf. "The Brain: An Orchestra Without a Conductor." *Max Planck Research* 3 (2005): 14–18.

5 *The Writings of Clement of Alexandria*. Edinburgh: T. & T. Clark, 1869.

6 Skakoon-Sparling, Shayna, Kenneth M. Cramer, and Paul A. Shuper. "The Impact of Sexual Arousal on Sexual Risk-Taking and Decision-Making in Men and Women." *Archives of Sexual Behavior* 45, no. 1 (2016): 33–42.

7 Schopenhauer, Arthur. *Essays and Aphorisms*. London: Penguin, 2004.

8 Poincaré, Henri. *L'Invention mathématique, conférence faite à l'Institut général psychologique*. Paris: Au siège de la Société, 1908.

9 Nisbett, Richard E., and Timothy D. Wilson. "Telling More Than We Can Know: Verbal Reports on Mental Processes." *Psychological Review* 84, no. 3 (1977): 231.

10 Easterbrook, Gregg. *The Progress Paradox*. New York: Random House, 2004.
11 MacDonald, George. *Unspoken Sermons, Vol. 1*. Library of Alexandria, 1867.
12 James, William. *The Varieties of Religious Experience: A Study in Human Nature; Being the Gifford Lectures on Natural Religion Delivered at Edinburgh in 1901–1902*. New York: Longmans, Green, 1911.
13 Lightman, Alan. *The Accidental Universe: The World You Thought You Knew*. New York: Knopf Doubleday, 2014.

SPIRITS EVERYWHERE

1 Crick, Francis, and J. Clark. "The Astonishing Hypothesis." *Journal of Consciousness Studies* 1, no. 1 (1994): 10–16.
2 Subbotsky, Eugene. "The Ghost in the Machine: Why and How the Belief in Magic Survives in the Rational Mind." *Human Development* 54, no. 3 (2011): 126–43.
3 Subbotsky, Eugene. *Magic and the Mind: Mechanisms, Functions, and Development of Magical Thinking and Behavior*. New York: Oxford University Press, 2010.
4 Zaitchik, Deborah, and Gregg E. A. Solomon. "Animist Thinking in the Elderly and in Patients with Alzheimer's Disease." *Cognitive Neuropsychology* 25, no. 1 (2008): 27–37.
5 Luu, Chi. "Personification Is Your Friend: The Language of Inanimate Objects." *JSTOR Daily*, 2016, https://daily.jstor.org/personification-is-your-friend-the-amazing-life-of-letters/.
6 Rojas-Rueda, D., et al. "Green Spaces and Mortality: A Systematic Review and Meta-analysis of Cohort Studies." *Lancet Planet Health* 3 (2019): 469–77.
7 Guthrie, Stewart Elliott, and Stewart Guthrie. *Faces in the Clouds: A New Theory of Religion*. Oxford: Oxford University Press on Demand, 1995.
8 Campbell, Joseph. *The Hero with a Thousand Faces*. New York: Pantheon Books, 1968.
9 Otto, Rudolf, and John Wilfred Harvey. *The Idea of the Holy. Translated by John W. Harvey. Fourth Impression, Revised with Additions*. Oxford: Oxford University Press, 1926.

10 James, William. *The Varieties of Religious Experience: A Study in Human Nature; Being the Gifford Lectures on Natural Religion Delivered at Edinburgh in 1901–1902*. New York: Longmans, Green, 1902.

11 Chesterton, Gilbert Keith. *Orthodoxy*. London: John Lane Company, 1908.

12 Barrett, Frederick S., Matthew W. Johnson, and Roland R. Griffiths. "Validation of the Revised Mystical Experience Questionnaire in Experimental Sessions with Psilocybin." *Journal of Psychopharmacology* 29, no. 11 (2015): 1182–90.

13 Yaden, David B., et al. "The Noetic Quality: A Multimethod Exploratory Study." *Psychology of Consciousness: Theory, Research, and Practice* 4, no. 1 (2017): 54.

14 Epley, Nicholas, Adam Waytz, and John T. Cacioppo. "On Seeing Human: A Three-Factor Theory of Anthropomorphism." *Psychological Review* 114, no. 4 (2007): 864.

15 Darwin, Charles. *The Descent of Man*. New York: D. Appleton & Company, 1871.

16 Broadhurst, P. L. *The Science of Animal Behaviour*. Baltimore: Penguin, 1963.

17 Hebb, D. O. "Emotion in Man and Animal: An Analysis of the Intuitive Processes of Recognition." *Psychological Review* 53 (1946): 88–106.

18 Dunbar, Robin I. M., and Susanne Shultz. "Evolution in the Social Brain." *Science* 317, no. 5843 (2007): 1344–7.

19 Holt-Lunstad, Julianne, et al. "Loneliness and Social Isolation as Risk Factors for Mortality: A Meta-analytic Review." *Perspectives on Psychological Science* 10, no. 2 (2015): 227–37.

20 Kanwisher, Nancy, Josh McDermott, and Marvin M. Chun. "The Fusiform Face Area: A Module in Human Extrastriate Cortex Specialized for Face Perception." *Journal of Neuroscience* 17, no. 11 (1997): 4302–11.

21 Gauthier, Isabel, et al. "Expertise for Cars and Birds Recruits Brain Areas Involved in Face Recognition." *Nature Neuroscience* 3, no. 2 (2000): 191.

22 Boden, Matthew Tyler. "Supernatural Beliefs: Considered Adaptive and Associated with Psychological Benefits." *Personality and Individual Differences* 86 (2015): 227–31.

23 DeLessio, Joe. "Why Superstitions Help Athletes Perform Better." *The Cut*, June 11, 2019, https://www.thecut.com/2015/06/why-superstitions-help-athletes-perform-better.html.

24 "Tom Brady Says Superstitious Gisele Bundchen Has Him Using 'Protection Stones'—and It Works." *CBS Boston*, 2019, https://boston.cbslocal.com/2019/02/07/tom-brady-gisele-bundchen-superstitions-super-bowl-stones/.

25 Damisch, Lysann, Barbara Stoberock, and Thomas Mussweiler. "Keep Your Fingers Crossed! How Superstition Improves Performance." *Psychological Science* 21, no. 7 (2010): 1014–20.

26 Barnes, A. "Man Fills Bowling Ball with His Father's Ashes as Tribute—and Bowls a Perfect Game." The Hill, April 23, 2021, https://thehill.com/changing-america/enrichment/arts-cultur e/549910-man-fills-bowling-ball-with-his-fathers-ashes-as.

THE UNCONSCIOUS IN THE LABORATORY

1 Bargh, John A., et al. "The Automated Will: Nonconscious Activation and Pursuit of Behavioral Goals." *Journal of Personality and Social Psychology* 81, no. 6 (2001): 1014.

2 Dijksterhuis, Ap, and Loran F. Nordgren. "A Theory of Unconscious Thought." *Perspectives on Psychological Science* 1, no. 2 (2006): 95–109.

3 Dijksterhuis, Ap, et al. "On Making the Right Choice: The Deliberation-Without-Attention Effect." *Science* 311, no. 5763 (2006): 1005–7.

4 Wilson, Timothy D., et al. "Introspecting About Reasons Can Reduce Post-choice Satisfaction." *Personality and Social Psychology Bulletin* 19, no. 3 (1993): 331–39.

5 Westen, Drew. "The Scientific Legacy of Sigmund Freud: Toward a Psychodynamically Informed Psychological Science." *Psychological Bulletin* 124, no. 3 (1998): 333.

6 Weinberger, Daniel A. "The Construct Validity of the Repressive Coping Style." In *Repression and Dissociation: Implications for Personality*

Theory, Psychopathology, and Health, edited by Jerome L. Singer, 337–86. Chicago: University of Chicago Press, 1990.

7 Freud, S. "On Beginning the Treatment." In *Standard Edition of the Complete Psychological Works of Sigmund Freud XII*. London: Hogarth Press, 1913, 135.

8 Evans, Jonathan St. B. T. "Dual-Processing Accounts of Reasoning, Judgment, and Social Cognition." *Annual Review of Psychology* 59 (2008): 255–78.

9 Schutte, Nicola S., et al. "Experiential and Rational Processing Styles, Emotional Intelligence and Wellbeing." *Australian Journal of Psychology* 62, no. 1 (2010): 14–19.

10 Jones, Garett, and W. Joel Schneider. "IQ in the Production Function: Evidence from Immigrant Earnings." *Economic Inquiry* 48, no. 3 (2010): 743–55.

11 Wolfradt, Uwe, et al. "Thinking Styles, Schizotypal Traits and Anomalous Experiences." *Personality and Individual Differences* 27, no. 5 (1999): 821–30.

12 Frazer, James George. *The Golden Bough*. London: Palgrave Macmillan, 1990, 701–11.

13 King, Laura A., et al. "Ghosts, UFOs, and Magic: Positive Affect and the Experiential System." *Journal of Personality and Social Psychology* 92, no. 5 (2007): 905.

14 Cheung, E. "Babe Ruth's 500th Home Run Baseball Bat Sold for $1 Million." *CNN*, December 17, 2019. www.cnn.com/2019/12/17/sport/babe-ruth-bat-baseball-sold-1-million-intl-hnk-scli/index.html.

15 Sobel, Dava. "Moon Dust." *New York Times*, October 1, 1995. https://www.nytimes.com/1995/10/01/magazine/hers-moon-dust.html.

16 King, Laura A., et al. "Ghosts, UFOs, and Magic: Positive Affect and the Experiential System." *Journal of Personality and Social Psychology* 92, no. 5 (2007): 905.

17 Ergle, Brittany, and Devin Puckett. "Superstitious Behavior and Locus of Control in Athletes, Musicians, and the General Population." *Showcase of Undergraduate Research and Creative Endeavors,* poster presentation, Winthrop University, 2015. https://digitalcommons.winthrop.edu/source/SOURCE_2015/posterpresentations/23/.

18 Risen, Jane L., and Thomas Gilovich. "Why People Are Reluctant to Tempt Fate." *Journal of Personality and Social Psychology* 95, no. 2 (2008): 293.

19 Epstein, Seymour. "Demystifying Intuition: What It Is, What It Does, and How It Does It." *Psychological Inquiry* 21, no. 4 (2010): 295–312.

20 Clarke, A. C. *Profiles of the Future*. New York: Holt, Rinehart, and Winston, 1984.

21 Weinberger, Adam B., et al. "Implicit Pattern Learning Predicts Individual Differences in Belief in God in the United States and Afghanistan." *Nature Communications* 11, no. 1 (2020): 1–12.

THE MAGICAL INSTINCT

1 Jung, C. G. "VIII Commentary on the Secret of the Golden Flower." In *Psyche and Symbol*, 333–98. Princeton, NJ: Princeton University Press, 2021.

2 Kahneman, Daniel. *Thinking, Fast and Slow*. New York: Farrar, Straus and Giroux, 2011.

3 Uvnäs-Moberg, Kerstin, et al. "Maternal Plasma Levels of Oxytocin During Physiological Childbirth—A Systematic Review with Implications for Uterine Contractions and Central Actions of Oxytocin." *BMC Pregnancy and Childbirth* 19, no. 1 (2019), article no. 285. https://bmcpregnancychildbirth.biomedcentral.com/articles/10.1186/s12884-019-2365-9.

4 Neumann, Erich. *The Origins and History of Consciousness*. New York: Routledge, 2015.

5 Bowlby, John. "The Nature of the Child's Tie to His Mother." *International Journal of Psycho-analysis* 39 (1958): 350–73.

6 Von Franz, Marie-Louise. *Archetypal Dimensions of the Psyche*. Boston: Shambhala, 1999.

7 Bekhrad, Joobin. "The Ancient Symbol That Spanned Millennia." *BBC*, December 4, 2017, https://www.bbc.com/culture/article/20171204-the-ancient-symbol-that-spanned-millennia.

8 Neumann, Erich. *The Origins and History of Consciousness*. New York: Routledge, 2015.

9 Proust, M. *Remembrance of Things Past. Vol. 1: Swann's Way: Within a Budding Grove*. Translated by C. K. Scott Moncrieff and Terence Kilmartin. New York: Vintage, 1982.

10 Rozin, Paul, Linda Millman, and Carol Nemeroff. "Operation of the Laws of Sympathetic Magic in Disgust and Other Domains." *Journal of Personality and Social Psychology* 50, no. 4 (1986): 703.

11 Miller, Peter. "A Thing or Two About Twins." *National Geographic*, January 2012. https://www.nationalgeographic.com/magazine/2012/01/identical-twins-science-dna-portraits/.

12 Schein, Elyse, and Paula Bernstein. *Identical Strangers: A Memoir of Twins Separated and Reunited*. New York: Random House, 2008.

13 Jung, Carl Gustav, et al. *Man and His Symbols*. New York: Doubleday, 1964.

14 Jung, C. G. *Collected Works of C.G. Jung, Volume 7: Two Essays in Analytical Psychology*. Princeton, NJ: Princeton University Press, 2014.

15 Von Franz, Marie-Louise. *Individuation in Fairy Tales: Revised Edition*. Boston: Shambhala, 2017.

16 Jung, Carl. *Psychological Types*. New York: Routledge, 2016.

17 Von Franz, Marie-Louise. *Individuation in Fairy Tales: Revised Edition*. Boston: Shambhala, 2017.

THE SHADOW

1 Schmeing, J.-B., et al. "Can the Neural Basis of Repression Be Studied in the MRI Scanner? New Insights from Two Free Association Paradigms." *PLoS ONE* 8, no. 4 (2013): e62358.

2 "Marvel's Iron Fist (TV Series)—Streaming Available." *Netflix Lovers USA*, 2019, https://www.netflixlovers.com/netflix-movies-and-tv-shows/80002612/marvels-iron-fist. Accessed 12/22/2019.

3 Jung, Carl. *Psychological Types*. New York: Routledge, 2016.

4 Levitz, Jennifer. "You Want 20% for Handing Me a Muffin? The Awkward Etiquette of iPad Tipping." *Wall Street Journal*, October 17, 2018. https://www.wsj.com/articles/you-want-20-for-handing-me-a-muffin-the-awkward-etiquette-of-ipad-tipping-1539790018.

5 Jung, C. G. "Aion: Researches into the Phenomenology of the Self." In *The Collected Works of C. G. Jung.* Vol. 9, Part 2. Princeton, NJ: Princeton University Press, 1959.

6 Schimel, Jeff, Jeff Greenberg, and Andy Martens. "Evidence That Projection of a Feared Trait Can Serve a Defensive Function." *Personality and Social Psychology Bulletin* 29, no. 8 (2003): 969–79.

FAIRY TALES

1 Zipes, Jack. *Breaking the Magic Spell: Radical Theories of Folk and Fairy Tales.* Lexington: University Press of Kentucky, 2002.

2 Winick, Stephen. "Einstein's Folklore." Folklife Today: American Folklife Center & Veterans History Project, Library of Congress, December 18, 2013. https://blogs.loc.gov/folklife/2013/12/einsteins-folklore/.

3 Dundes, Alan. *Cinderella, a Casebook.* Madison: University of Wisconsin Press, 1988.

4 Spake, Garrett. "How the Texas Chain Saw Massacre Was Inspired by a Brothers Grimm Story." *Screenrant,* May 24, 2020. https://screenrant.com/texas-chainsaw-massacre-movie-hansel-gretel-story-inspiration/.

5 de Saint-Exupéry, Antoine. *Le petit prince [The little prince].* Translated by Richard Howard. New York: Harvest, 1943.

6 Vitz, Paul C. "The Use of Stories in Moral Development: New Psychological Reasons for an Old Education Method." *American Psychologist* 45, no. 6 (1990): 709.

7 Jacobi, Jolande. *Complex/Archetype/Symbol in the Psychology of CG Jung.* New York: Routledge, 2013.

8 Jung, Carl Gustav. *Aion: Researches into the Phenomenology of the Self.* New York: Routledge, 2014.

9 Gallese, Vittorio. "Embodied Simulation Theory: Imagination and Narrative." *Neuropsychoanalysis* 13, no. 2 (2011): 196–200.

10 Stephens, Greg J., Lauren J. Silbert, and Uri Hasson. "Speaker–Listener Neural Coupling Underlies Successful Communication." *Proceedings of the National Academy of Sciences* 107, no. 32 (2010): 14,425–30.

11 Kidd, David Comer, and Emanuele Castano. "Reading Literary Fiction Improves Theory of Mind." *Science* (2013): 377.

12 Johnson, Dan R. "Transportation into a Story Increases Empathy, Prosocial Behavior, and Perceptual Bias Toward Fearful Expressions." *Personality and Individual Differences* 52, no. 2 (2012): 150–55.

13 Barraza, Jorgea, and Paulj Zak. "Empathy Toward Strangers Triggers Oxytocin Release and Subsequent Generosity." *Annals of the New York Academy of Sciences* 1167, no. 1 (2009): 182–89.

14 Sturm, Brian W. "The Enchanted Imagination: Storytelling's Power to Entrance Listeners." *School Library Media Research* 2, no. 6 (1999): 1–21.

15 Jin, Putai. "Efficacy of Tai Chi, Brisk Walking, Meditation, and Reading in Reducing Mental and Emotional Stress." *Journal of Psychosomatic Research* 36, no. 4 (1992): 361–70.

16 Gleed, Alasdair. *Booktrust Reading Habits Survey 2013*. Stockport, UK: DJS Research Limited, 2013.

17 Lutz, Cora E. "The Oldest Library Motto: ψγxhσ Iatpeion." *The Library Quarterly* 48, no. 1 (1978): 36–39.

18 Subbotsky, Eugene, Claire Hysted, and Nicola Jones. "Watching Films with Magical Content Facilitates Creativity in Children." *Perceptual and Motor Skills* 111, no. 1 (2010): 261–77.

19 Plucker, Jonathan A. "Is the Proof in the Pudding? Reanalyses of Torrance's (1958 to Present) Longitudinal Data." *Creativity Research Journal* 12, no. 2 (1999): 103–14.

ALCHEMY

1 Valentine, Basil. "Triumphal Chariot of Antimony," in Lawrence Principe, *The Secrets of Alchemy*. Chicago: University of Chicago Press, 2013.

2 Jung, Carl Gustav. *Aspects of the Feminine.* Princeton, NJ: Princeton University Press, 1982.

3 Principe, Lawrence. *The Secrets of Alchemy*. Chicago: University of Chicago Press, 2013.

4 Von Franz, Marie-Louise. *Alchemy: An Introduction to the Symbolism and the Psychology, Volume 5*. Scarborough, Ontario: Inner City Books, 1980.

5 Kövecses, Zoltán. *Emotion Concepts*. Berlin: Springer Science & Business Media, 2012.

6 Omori, Ayako. "Emotion as a Huge Mass of Moving Water." *Metaphor and Symbol* 23, no. 2 (2008): 130–46.

7 Von Franz, Marie-Louise. *Alchemy: An Introduction to the Symbolism and the Psychology, Volume 5*. Scarborough, Ontario: Inner City Books, 1980.

8 Jung, Carl Gustav. *Psychology of the Transference.* In *The Collected Works of C. G. Jung, Vol. 16.* Princeton, NJ: Princeton University Press, 2020.

9 Jung, Carl Gustav. *Psychology and Alchemy*. New York: Routledge, 2014.

MYSTICAL NUMBERS

1 Wigner, E. "The Unreasonable Effectiveness of Mathematics in the Natural Sciences." *Communications in Pure and Applied Mathematics* 13, no. 1 (1960): 1–14.

2 Cliff, Harry. "Beyond the Higgs: What's Next for the LHC?" The Royal Institution, January 17, 2018. YouTube video, 59:44. https://youtu.be/edvdzh9Pggg.

3 Augustine. *The City of God*. Peabody, MA: Hendrickson Publishers, 2009.

4 MacDonald, George. *Unspoken Sermons, Vol. 1*. Library of Alexandria, 1867.

5 Jung, Carl Gustav. *The Archetypes and the Collective Unconscious*. Translated by R. F. C. Hull. Princeton, NJ: Princeton University Press, 1959.

6 Macrobius, Ambrosius Aurelius Theodosius. *Commentary on the Dream of Scipio, Vol. 48*. New York: Columbia University Press, 1990.

7 Wikipedia. "Tetractys." Last modified May 5, 2020. https://en.wikipedia.org/wiki/Tetractys.

8 Dantzig, Tobias. *Number: The Language of Science*. London: Plume, 2007.

THE TAROT

1 Place, Robert. *The Tarot: History, Symbolism, and Divination*. New York: Penguin, 2005.

2 Jones, Josh. "Why Butt Trumpets & Other Bizarre Images Appeared in Illuminated Medieval Manuscripts." *Open Culture*, December 8, 2020. https://www.openculture.com/?p=1084440.

3 Ray, Sharmistha. "Reviving a Forgotten Artist of the Occult." *Hyperallergic*, March 23, 2019. https://hyperallergic.com/490918/pamela-colman-smith-pratt-institute-libraries/.

4 Place, Robert. *The Tarot: History, Symbolism, and Divination*. New York: Penguin, 2005.

BECOMING TRANSCENDENT

1 C. Robert Cloninger, et al. *The Temperament and Character Inventory (TCI): A Guide to Its Development and Use*. St. Louis, MO: Center for Psychobiology of Personality, Washington University, 1994.

2 Levenson, Michael R., et al. "Self-Transcendence: Conceptualization and Measurement." *The International Journal of Aging and Human Development* 60, no. 2 (2005): 127–43.

3 Costa, Paul T., and Robert R. McCrae. *NEO Personality Inventory–Revised (NEO PI–R)*. Odessa, FL: Psychological Assessment Resources, 1992.

4 Jung, C. G. "The Transcendent Function." In *The Collected Works of C. G. Jung, Vol. 16*. Princeton, NJ: Princeton University Press, 1969.

5 Wikipedia. "Éliphas Lévi." Last modified May 26, 2020. en.wikipedia.org/wiki/Éliphas_Lévi.

6 Jung, C. G. *Collected Works of C. G. Jung, Vol. 7: Two Essays in Analytical Psychology*. Princeton, NJ: Princeton University Press, 2014.

7 Tornstam, Lars. "Maturing into Gerotranscendence." *Journal of Transpersonal Psychology* 43, no. 2 (2011): 166–180.

8 Killingsworth, Matthew A., and Daniel T. Gilbert. "A Wandering Mind Is an Unhappy Mind." *Science* 330, no. 6006 (2010): 932.

9 Jung, Carl Gustav. *Memories, Dreams, Reflections*. New York: Vintage, 1989.

10 Hannah, Barbara. "Some Remarks on Active Imagination." *Spring* (1953): 38–58.

11 Dallett, Janet. "Active Imagination in Practice." *Carl Gustav Jung: Psychopathology and Psychotherapy* (1992): 235.

THE *CIRCULATIO* AND THE *CONJUNCTIO*

1 Orme-Johnson, David. "Medical Care Utilization and the Transcendental Meditation Program." *Psychosomatic Medicine* 49, no. 5 (1987): 493–507.

2 Anālayo, Bhikkhu. "The Luminous Mind in Theravāda and Dharmaguptaka Discourses." *Journal of the Oxford Centre for Buddhist Studies* 13 (2017): 10–51.

3 Mahony, William K. *The Artful Universe: An Introduction to the Vedic Religious Imagination*. Albany: State University of New York Press, 1998.

4 *Rig Veda Upanishad* 7.33.9.

5 Jung, Carl Gustav. *Commentary on "The Secret of the Golden Flower."* Princeton, NJ: Princeton University Press, 2020.

6 Ricard, Matthieu, Antoine Lutz, and Richard J. Davidson. "Mind of the Meditator." *Scientific American* 311, no. 5 (2014): 38–45.

7 Lazar, Sara W., et al. "Meditation Experience Is Associated with Increased Cortical Thickness." *Neuroreport* 16, no. 17 (2005): 1893.

8 McCracken, Lance M., Kevin E. Vowles, and Christopher Eccleston. "Acceptance of Chronic Pain: Component Analysis and a Revised Assessment Method." *Pain* 107, no. 1–2 (2004): 159–66.

9 Ibid.

10 Jung, C. G. "Aspects of the Feminine". In *Collected Works of C.G. Jung, Vols. 6, 7, 9i, 9ii, 10, 17*). Princeton, NJ: Princeton University Press, 2020.

11 Wikipedia. "True Name." Last modified July 24, 2021. en.wikipedia.org/wiki/True_name.

12 Jung, Carl Gustav. *Symbols of Transformation*. New York: Routledge, 2014.

13 Tolkien, J. R. R. *The Lord of the Rings: The Two Towers*. London: George Allen & Unwin, 1954.

14 Jansson, Tove. *Tales from Moominvalley*. New York: Farrar, Straus and Giroux, 2014.

15 Nidich, Sanford, et al. "Non-trauma-focused Meditation Versus Exposure Therapy in Veterans with Post-traumatic Stress Disorder: A Randomised Controlled Trial." *The Lancet Psychiatry* 5, no. 12 (2018): 975–86.

16 Beru Khyentze Rinpoche. *The Mahāmudrā Eliminating the Darkness of Ignorance*. Library of Tibetan Works and Archives, 1989.

17 Von Franz, Marie-Louise. *Archetypal Dimensions of the Psyche*. Boston: Shambhala, 1999.

18 Lipscombe, Neil. "The Relevance of the Peak Experience to Continued Skydiving Participation: A Qualitative Approach to Assessing Motivations." *Leisure Studies* 18, no. 4 (1999): 267–88.

19 Schlosser, Marco, et al. "Unpleasant Meditation-Related Experiences in Regular Meditators: Prevalence, Predictors, and Conceptual Considerations." *PloS ONE* 14, no. 5 (2019): e0216643.

20 Cebolla, Ausiàs, et al. "Unwanted Effects: Is There a Negative Side of Meditation? A Multicentre Survey." *PloS ONE* 12, no. 9 (2017): e0183137.

21 Cicero, Marcus Tullius. *De Officiis*. Cambridge: Harvard University Press, 1913.

22 Crisp, Roger, ed. *Aristotle: Nicomachean Ethics*. Cambridge, UK: Cambridge University Press, 2014.

23 Lockwood, Thornton C. "Habituation, Habit, and Character in Aristotle's Nicomachean Ethics." *A History of Habit: From Aristotle to Bourdieu* 19 (2013): 36.

24 Bowlby, J. *A Secure Base: Parent–Child Attachment and Healthy Human Development*. New York: Basic Books, 1988.

25 Shahar, Ben, et al. "A Wait-List Randomized Controlled Trial of Loving-Kindness Meditation Programme for Self-criticism." *Clinical Psychology & Psychotherapy* 22, no. 4 (2015): 346–56.

26 Rector, Neil A., et al. "Self-criticism and Dependency in Depressed Patients Treated with Cognitive Therapy or Pharmacotherapy." *Cognitive Therapy and Research* 24, no. 5 (2000): 571–84.

27 Fredrickson, Barbara L., et al. "Open Hearts Build Lives: Positive Emotions, Induced Through Loving-Kindness Meditation, Build Consequential Personal Resources." *Journal of Personality and Social Psychology* 95, no. 5 (2008): 1045.

28 Aquinas, Thomas. *Summa Theologica*, II–II, q. 25, a. 4.

29 Lutz, Antoine, et al. "Regulation of the Neural Circuitry of Emotion by Compassion Meditation: Effects of Meditative Expertise." *PloS ONE* 3, no. 3 (2008): e1897.

30 Condon, Paul, et al. "Meditation Increases Compassionate Responses to Suffering." *Psychological Science* 24, no. 10 (2013): 2125–27.

THE KEYS OF THE TAROT AND THE PERFECTION OF THE SOUL

1 "Death Angel Drummer: I Visited Hell, Saw Satan During COVID-19 Coma." May 21, 2020. Loudwire, 1:33. https://www.youtube.com/watch?v=aMlskeZoL5w; Lapin, Tamar. "Death Angel Drummer Says He Met Satan During Near-Death Coronavirus Coma." *New York Post*, May 19, 2020. https://nypost.com/2020/05/19/death-angel-drummer-met-satan-during-near-death-coronavirus-coma/.

2 McGaughy, Lauren. "Texas Lawmaker Caught on Video Dropping Envelope of Cocaine Turns Himself into Authorities." *Dallas News*, November 14, 2019. https://www.dallasnews.com/news/politics/2019/11/14/texas-lawmaker-caught-on-video-dropping-envelope-of-cocaine-admits-hes-stepping-down-due-to-addiction/.

3 Jung, Carl, et al. *Man and His Symbols*. New York: Doubleday, 1964, p. 170.

4 Gurshtein, Alex A. "Did the Pre-Indo-Europeans Influence the Formation of the Western Zodiac?" *Journal of Indoeuropean Studies* 33, nos. 1/2 (2005): 103.

5 Wikipedia. "Anima Mundi." Last modified May 5, 2021. https://en.wikipedia.org/wiki/Anima_mundi.

INDEX

N

T

U

V

W

Y

ABOUT THE AUTHOR

Daniel Z. Lieberman, MD, is professor of psychiatry and behavioral sciences at George Washington University and the coauthor of *The Molecule of More: How a Single Chemical in Your Brain Drives Love, Sex, and Creativity—and Will Determine the Fate of the Human Race*. Dr. Lieberman is a distinguished fellow of the American Psychiatric Association, a recipient of the Caron Foundation Research Award, and has published more than fifty scientific reports in the areas of brain science and human behavior. He has provided insight on psychiatric issues for the US Department of Health and Human Services, the US Department of Commerce, and the Office of Drug and Alcohol Policy and Compliance, and is a frequent guest on radio, television, and podcasts. Dr. Lieberman studied the Great Books at St. John's College. He received his medical degree and completed his psychiatric training at New York University.